WWII U.S. Sherman Tank

In Action®

Written by Rob Ervin
and David Doyle

Cover Art by Don Greer

Line Illustrations by Melinda Turnage

Squadron/Signal Publications

(Front Cover) A 5th Armored Division M4A3 traverses a dusty trail through the fall colors of Fort Knox, Kentucky, in 1943. Crews trained at various locations in the U.S. and Britain for combat in North Africa, Europe, and the Pacific. Units received new tanks once they were shipped to their theater of operation.

(Back Cover) The crew of *Bushmaster,* an Army M4 composite-hull Sherman of the 763rd tank Battalion, faces a tough job to extract their tank from the drying mud of Leyte in November 1944. The Army units in the Pacific often painted out the large white stars on the turret sides and hull in an attempt to better camouflage their tanks.

About the In Action® Series

In Action® books, despite the title of the genre, are books that trace the development of a single type of aircraft, armored vehicle, or ship from prototype to the final production variant. Experimental or "one-off" variants can also be included. Our first *In Action*® book was printed in 1971.

ISBN 978-0-89747-630-0

Proudly printed in the U.S.A.

1115 Crowley Drive, Carrollton, TX 75006-1312 U.S.A.

Military/Combat Photographs and Snapshots

If you have any photos of aircraft, armor, soldiers, or ships of any nation, particularly wartime snapshots, why not share them with us and help make Squadron/Signal's books all the more interesting and complete in the future? Any photograph sent to us will be copied and returned. Electronic images are preferred. The donor will be fully credited for any photos used. Please send them to:

Squadron/Signal Publications
1115 Crowley Drive
Carrollton, TX 75006-1312 U.S.A.
www.SquadronSignalPublications.com

(Title Page) A new M4A4 Sherman is unloaded at Casablanca Harbor, Morocco, in 1943. The tank is part of a Lend-Lease shipment for Free French Forces in North Africa. The tank is sealed for its sea voyage with mastic. (National Archives)

Acknowledgments

Special thanks to Mike Canaday and Kurt Laughlin for tolerating my countless questions and for sharing their vast knowledge on this subject. My work on this project would not have been possible without the kind support and encouragement of my wife, Linda. *Rob Ervin*

This book would not have been possible without the generous assistance of Joe DeMarco; Claude Gillono; Scott Taylor; Steve Zaloga; the staff of the Patton Museum; the National Archives; the TACOM LCMC History Office; and the Naval History and Heritage Command. Special thanks to my wife Denise, who patiently traveled with me coast to coast as we researched this volume. *David Doyle*

Introduction

The M4 Medium Tank was the Western Allies' most widely produced tank during World War II. Its cumulative output of nearly 50,000 units far exceeded Germany's entire tank production from 1939 to 1945. With tens of thousands of Shermans in the supply lines, the M4 became the major tank type used by the Western Allies, seeing combat under an array of flags on every front. Its complete story would far exceed the limits of this volume. Accordingly, we will focus on the component changes among the variants and sub-variants of the Shermans used by the U.S. forces during WWII.

As the clouds of war gathered over Europe, the Ordnance Department in the U.S. contracted for a production run of 1,000 M2A1 Medium Tanks. But the armored engagements that preceded the fall of France in 1940 revealed that the M2A1 would be obsolete before it reached the field. The contract was therefore shifted to the M3 Medium, which had yet to be designed, as a stop-gap measure, pending development of a tank with a cast turret that would accept the 75mm gun and traverse 360°. The "General Lee," as the M3 would come to be known, had its 75mm gun mounted in a sponson with limited traverse, while secondary 37mm armament was fitted in the turret. With riveted construction and a tall silhouette due to its radial engine, the M3 design was not ideal.

The M4 would carry over the lower hull suspension and drive train from the M3. The hull was to be a single large casting. The turret would have a ring of 69 inches and a removable front plate to allow a change of armament. The suspension and tracks were similar to the type proven on the U.S. light tanks of the late 1930s.

The tank was conceived as an infantry support weapon, with a dual-purpose gun. Its role was to exploit a breakthrough created by infantry and artillery assault and wreak havoc in the enemy's rear area. The doctrine of the time considered enemy tanks to be the responsibility of anti-tank guns and purpose-built tank destroyers in units supporting the infantry divisions. In reality this doctrine was quite flawed, and it slowed future efforts to develop the M4 series as a more capable tank killer.

Even after production began in 1942, the design continued to change almost constantly as new components were introduced into the assembly plants. Lessons learned in combat and during manufacture, plus feedback from troops in the field helped to bring about changes in the tanks produced. A search for new power plants brought about the use of a total of five engines. The M4 series was built with three different main guns. It is a tribute to the design that production changes in weapons, suspension, power trains, and armor could be introduced without stopping production. The M4 Sherman tanks being produced in 1945 bore little resemblance to those coming off assembly lines in 1942.

The many variations in Sherman production can be categorized in various ways. One popular classification, which is used in this volume, pertains to hull hatch configuration. "Small hatch" tanks sport hatches that swing open parallel to the hull sides, and are hinged at two points. The later, "large hatch" tanks have hatches with a single hinge point, that when open are not parallel to the hull sides.

The Sherman was a simple, reliable, and flexible design that fit well the manufacturing capabilities of the United States.

The M3 Medium Tank, known as the Lee, was merely a stop gap measure. The riveted hull and sponson mount for the 75mm gun were shortcomings. The design served to train crews and equip Allied armies until a design with a 360° traversing turret could be made available. (Patton Museum)

The T6 served as the prototype for the M4 series. It was developed on the running gear and lower hull of the M3 Medium tank. The counterweights on the gun barrel were necessary to balance the short M2 75mm gun, which was to be replaced by the longer M3 when available. (Patton Museum)

Production Begins

The initial M4 production tanks rolled off the assembly line at Lima Locomotive Works in February of 1942. The first tanks built were M4A1s with cast hulls and had a number of detail differences from the prototype T6, such as the elimination of hull side doors and the addition of a hatch in the hull roof above the co-driver.

A total of nine industrial plants were eventually involved in the assembly of the Sherman tanks from castings, engines, weapons, and steel shipped to them from all over the United States. Production followed basically the assembly-line pattern established by the American automobile industry, and the firms involved in turning out the tanks included Chrysler, Ford, and General Motors, as well as several locomotive manufacturers. The huge Chrysler Tank Arsenal in Detroit alone eventually claimed assembly of over 22,000 medium and heavy tanks.

At the time that production began, American tank units were in combat only in the Pacific. The new tanks were shipped to U.S. Armored Divisions training within the United States and to the British fighting in North Africa. The influx of the new M4, along with the M3 Mediums, helped the British turn the tide of battle against Erwin Rommel's Deutsches Afrikakorps. In July of 1942, 300 new Shermans and 100 M7 Priests were pulled from U.S. units and shipped to Egypt to make good the British losses suffered during the spring battles. By the time the Germans attacked al-'Alamayn ("El Alamein") in October 1942, British forces were well equipped with new M4A1s and M4A2s.

The earliest production examples of M4A1 differed from the T6 in several key details. The hull side doors were eliminated, and a hatch was added to the roof above the co-driver's head. The tanks retained the twin fixed front-mounted Browning .30-caliber machine guns. (Patton Museum)

M4A1 Production began at the Lima Locomotive Works in Lima, Ohio, in February 1942. Here very early tanks with gun sight rotors in the turret roof and M3 style exhausts await final assembly. (Joe DeMarco collection)

The second production tank was shipped to the British and was displayed in London. Named "Michael" in honor of the head of the British Tank Mission, Michael Dewar, the tank remains on display at the Tank Museum, Bovington. (Patton Museum)

The M4, although first in designation, was actually preceded in production by the M4A1. Both were powered by the R975 air-cooled radial engine. The earliest models of the Sherman can be characterized by the 'small' hatches for driver and co-driver, which open parallel to the hull sides. (National Archives)

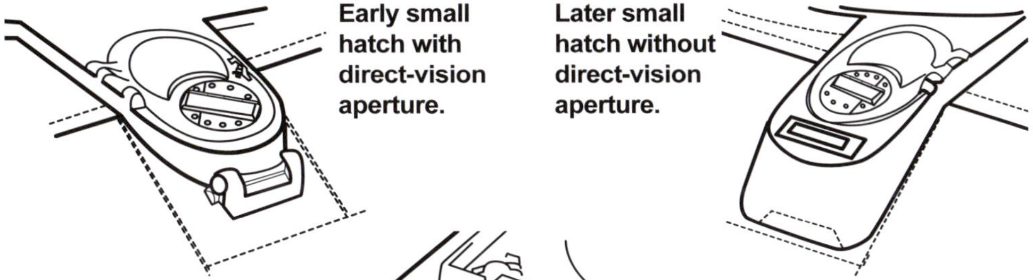

Early small hatch with direct-vision aperture.

Later small hatch without direct-vision aperture.

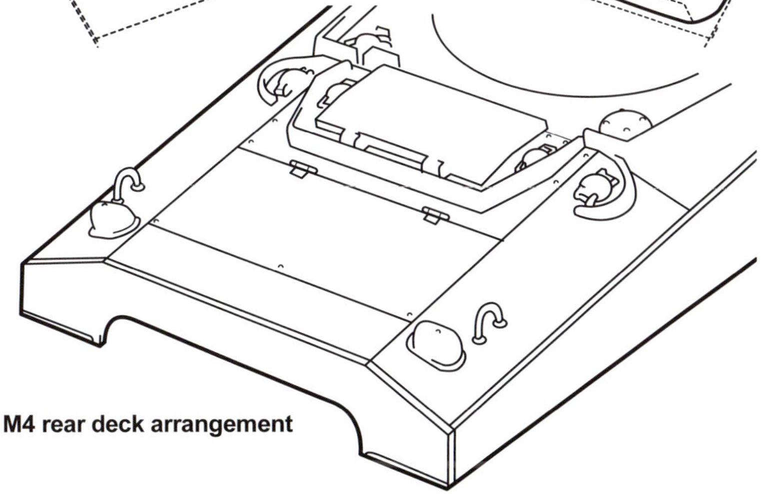

M4 rear deck arrangement

Small-Hatch M4

Although the name M4 might suggest that this was the initial and definitive design of the M4 medium tank series, in fact the design for the welded-hull tank was an attempt to decrease the anticipated bottleneck in production of the M4A1's large, cast upper hull. The M4 went into production at Pressed Steel Car Company in July 1942, following the M4A1 by almost five months. The M4 and M4A1 tanks were mechanically identical and were powered by the Continental air-cooled R975 engine. The welded hull provided room for the storage of an additional seven rounds of 75mm ammunition. When production ended in January 1944, a total of 6,748 75mm M4s had been turned out by five manufacturers.

Power was transferred to the front of the vehicle through a prop shaft to a five-speed syncromesh transmission and final-drive assembly. The radial engine dictated that this shaft pass through the fighting compartment quite high, elevating the height necessary for the floor of the turret basket and giving the tank a very tall silhouette.

The crew consisted of five men: driver and co-driver/front gunner in the hull front; the gunner, loader, and commander in the turret. The turret was equipped with a large rotating cupola for the commander, with a mount for a .50-caliber anti-aircraft machine gun. The single hatch proved inadequate as an emergency exit for the three turret crew members, as were the small hull hatches.

Initial production examples were built with the suspension carried over from the M3 Medium Tank series, with the three-piece housing for the differential and final drive and the track return support roller centered above the bogie. The turret was the low-bustle type with the narrow M34 gun mount that used a link to the gunner's sight mounted in the right side turret roof.

Very quickly production details began to change as components were improved. A single-piece cast differential cover was introduced. Ears appeared on the outer shield of the M34 gun mount to protect the recoil area of the gun barrel. A heavy-duty suspension was introduced with return roller to the rear of the bogie, and driver's hood castings without the direct-vision doors were installed. It is difficult to establish an exact time line for the introduction of these improvements into production models, since many manufacturers were involved, and since there was a desire to use up stocks of older components. As an example, all remaining direct-vision driver's hoods were gathered and shipped to Baldwin Locomotive, which used them until the end of M4 75mm production. On later vehicles these doors were welded shut and covered by an added appliqué plate at the factory. These practices – along with upgrades made at the unit or depot level, such as the addition of appliqué armor or replacement of the M34 gun mount with the new wide M34A1 with a direct sight telescope for the gunner – make it difficult to ascertain a particular tank's production period based only on photographic images of the vehicle.

The M4 with the 75mm gun along with the M4A1 were the only tanks used by U.S. forces through the landings in North Africa, Sicily, Italy, and France up until the late summer of 1944. These tanks were seen in decreasing numbers in armored divisions right through the end of the war in Europe.

An M4 tank undergoes winter tests along the Alaska Highway on 9 February 1944. The tank is equipped with a full set of sand shields and hull appliqué armor plates. The device to the left of the turret hatch is a clear cylindrical hood that allowed the commander full vision while being protected from the cold. (TACOM LCMC History Office)

The Sherman first saw combat in U.S. hands with the 1st Armored Division south of Sayyidî Bû Zayd ("Sidi Bou Zid"), Tunisia, on 14-15 February 1943. This early direct-vision hull tank, featuring M3-style suspension, the E4186 single-piece differential cover, and early M34 gun shield, was disabled by a Panzer III's 5cm rounds. The 1st Armored Division had a few welded-hull M4s, but most of its Shermans were cast-hull M4A1s.

An M4 of the 751st Tank Battalion comes ashore at Anzio. The tank is completely prepared for deep-water fording. The areas around the vision doors, front machine gun, and main gun have been sealed with waterproofing mastic. The deep-wading stacks manufactured in North Africa have a distinctive capped mushroom shape and differ greatly from those used later in France. There is a dark tide mark on the front of tank's differential cover. (Patton Museum)

Baldwin-built M4 tanks of the 1st Armored Division, 13th Tank Battalion land from an LST at Anzio in late April 1944. Lengths of track have been added to the front of the tanks for additional armor protection. These tanks are still equipped with direct-vision doors in the driver's hoods, yet have the later wide rotor shields of the M34A1 gun mount. (Patton Museum)

In preparation for the Normandy landings wading stacks are installed on tanks in England on 5 April 1944. The stacks allowed the tanks to wade ashore from landing craft without swamping the engines. The devices were designed with a quick release to jettison the upper portions inland. The M4 is equipped with a full set of appliqué armor. In the background is an M10 Tank Destroyer. (National Archives)

An M4 of the 755th Tank Battalion advances near Acquafondata, Italy, on 12 January 1944. The rear of the vehicle is marked with the Allied star with surround, and has round air cleaners and the remains of sand shields. The turret is the early low-bustle design with a pistol port in the left side. (National Archives)

Southwest England bristled with activity and military vehicles in the spring of 1944. A 6th Armored Division Sherman in full markings climbs a village street. All American M4s were fitted with appliqué plates and the wide M34A1 gun mount in English workshops prior to the D-Day landings. (National Archives)

M4s of the 757th TB ride through St. Giorgio, Italy, on 16 May 1944, after supporting the Free French forces' attack on the town. The M3A1 White Scout car bears French markings. The large stars with surrounds on the M4 turrets were common in the Italian campaign and helped to avoid confusion of stars with German crosses at great distances. Behind the first M4 is a T2 recovery vehicle based on the M3 Lee. (National Archives)

The Bocage country of Normandy was difficult tank terrain. The narrow roads were lined with thick centuries-old hedgerows that divided the farmer's fields. These provided the German defenders excellent hiding places for machine guns and tank destroyers. Here a Sherman advances along a narrow road in support of infantry. (Patton Museum)

Hedgerow cutting devices were added to the differential cover of Sherman tanks to allow them to drive through the thick foliage without exposing the thin armor of the hull floor. These devices were first developed by Sgt. Curtis Culin, and were produced in a number of variations from the steel of German beach obstacles. (Patton Museum)

M4s face only minimal resistance as they enter the port city of Cherbourg, France, on 27 June 1944. Allied naval and aerial bombardment had laid waste to the town during the first few weeks following the Normandy landings, and on 26 June, the 79th Infantry Division had taken the local garrison. (National Archives)

Both the M4 and the M4A1 are powered by the Continental R975-C1 9-cylinder, 4-cycle, air-cooled radial. The engine's aircraft lineage is readily apparent, and it was that lineage that led in part to production bottlenecks, due to the overwhelming wartime demands for both aircraft and tank engines.

VVSS Road Wheel Variations

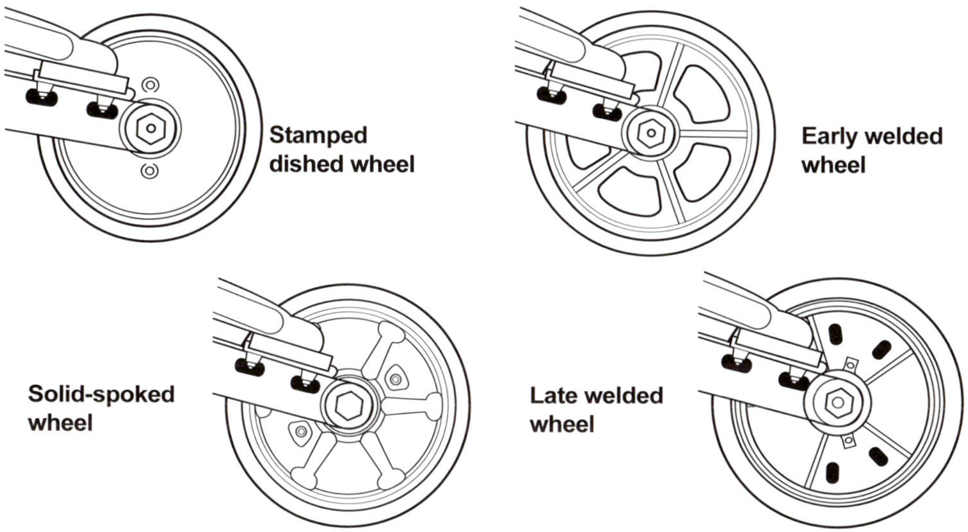

U.S. tank crews soon learned that the armor of the M4 was very vulnerable to the German anti-tank weapons they faced. Many different forms of additional armor were tried to give some measure of added protection. Sandbags, often filled with cement, were applied to the glacis of many tanks to give some stand off protection against shaped charge weapons such as the Panzerfaust and Panzerschreck. (Patton Museum)

An M4 crosses the Seine River on a pontoon bridge on 28 August 1944. By this time many 1st Army units had adopted bands of black disruptive camouflage paint over the tank's Olive Drab finish, often painting out white star markings. Engineer units were crucial to facilitating the Allied advance across France. (U.S. Army Engineer School History Office)

An M4 crossing an M2 treadway bridge displays logs applied to the hull sides and turret for additional armor protection. Shaped charge weapons used chemical energy to burn through armor plate. It was hoped that a round striking a log would expend its energy before penetrating the tank's armor. (Patton Museum)

M4s of the 4th Armored Division advance across a snow covered field near Bastogne on 3 January 1945. The 4th Armored Division was the first unit to reach the besieged airborne units cut off in Belgium during the battle of the Bulge. By this time the 75mm M4s were being displaced in most units by newer M4A3s and 76mm-armed tanks. (National Archives)

A 6th Armored Division M4 with well-worn whitewash camouflage carries a group of Armored Infantrymen, amply equipped in snow-pac boots and overcoats. The hull side carries the unique hull numbers of the Division. The tank's tracks are equipped with extended end connectors, often called "duck bills." These end connectors were added to the track to decrease the ground pressure exerted by the vehicle and enhance cross-country performance. (Patton Museum)

Many Shermans used in training were sent back through various facilities for re-manufacturing. This continuing program brought the older tanks up to the latest standard. This M4, photographed in October of 1944, has received a full set of appliqué armor, a gun travel lock, relocation of the siren to the front armor behind a guard, and short fender extensions to allow the use of the sand shields with the extended track end connectors. (Patton Museum)

An internal ammunition explosion has blown out this M4's sponson floor plate, breaking the left side track. Appliqué plates were added over areas where the ammunition racks were located in the sponsons, in an attempt to reduce the likelihood of penetration in these vulnerable areas. Even minor penetrations could set off fires fueled by the propellant of the tank's stowed rounds. (Patton Museum)

A T2 recovery vehicle prepares to retrieve an out-of-commission M4 in Aldorf Germany on 15 October 1944. The tracks are being removed to facilitate towing the tank. U.S. armored units were equipped with excellent recovery facilities, and unit workshops labored round the clock to have the recovered tanks repaired and returned to service. (National Archives)

This was an all too common sight, a burned M4 and a dead crewman. The Sherman gained a reputation for inadequate protection and acquired the nickname Ronson, after the famous cigarette lighter, because of being prone to catch fire once penetrated. The addition of appliqué armor plates was the first official step to help alleviate these concerns. (National Archives)

A whitewashed M4 leads a column of soft-skinned vehicles down a tree lined road. The white camouflage paint was applied to avoid the stark contrast between the dark olive-drab vehicles and the snowy terrain. Often the color was applied with a mixture of lime, salt, and water when sufficient white paint was unavailable. (Don F. Pratt Museum)

An M1 wrecker has pulled the engine of *Hurricane,* an M4 of 2nd Armored Division, for replacement. The M4 was powered by the Continental air-cooled R975-C1 radial engine, a development of an aircraft engine. The shiny new replacement engine lies on the ground awaiting installation. (National Archives)

Lucky Legs II, an M4 of the 754th Tank Battalion, advances covering infantry in the jungle of Bougainville during March of 1944. The M4 far outclassed any Japanese armor that it faced in the Pacific campaigns. Close co-operation with infantry helped to protect the tanks from attack by enemy troops hidden in the thick jungle. (Department of Defense)

This later-production M4 Sherman is equipped with the E8543 double brake differential assembly that can be identified by its more pointed cross section. The tank is operating in the Bougainville jungle in early 1944 and retains the large national star markings. The vehicle in the background is an M3 Light Tank. (National Archives)

The turret of this M4 has many earmarks of early-production vehicles. The M34 gun mount has the early narrow rotor shield without the protective ears for the recoil area of the gun barrel. The pintle mount for the .50-caliber Browning anti-aircraft machine gun is the early two position type, and the clip to hold the early 50-round ammunition box is evident on the side of the gun. (National Archives)

The Autobahn, one of the supreme technological achievements of the German Reich, also served as a tool of its conquerors. U.S. Army M4s race into the heart of Germany on the highway. The foreground vehicle features a sheet metal exhaust deflector, “duck bill” extended end connectors, and an all-round vision cupola. (National Archives)

Small-Hatch M4A1

The Pacific Car and Foundry pilot model M4A1 tank is seen at the Aberdeen Proving Grounds in Maryland on 9 June 1942. Visible on this vehicle are all the features typical of early standardized production M4A1s. These early features include the M3-style suspension with the return roller centered above the bogie assembly. The vehicle's tracks are the early reversible T41 rubber block style, which was soon replaced by the similar T51 non-reversible type of track. The hull features direct-vision doors, and the M34 gun mount has the early narrow rotor shield. The rotor over the gunner's sight has here been replaced with an M4 periscope gun sight. The mounts for the two fixed machine guns in the lower front hull are covered by tape. These machine guns would soon be eliminated and on some tanks the hull openings would be plugged. (National Archives)

The first tank to roll off the Lima Locomotive Works assembly line in February 1942 was the M4A1. The initial production batch differed from later models by having two fixed .30-caliber Browning machine guns in the lower front hull, and a rotating shield on the right turret roof for the gunner's periscope sight. Soon these features were eliminated by the standard production type that featured the low-bustle turret with M34 gun mount for the M3 75mm gun. Early tanks retained the riveted construction of the lower hull, plus the suspension, tracks, and Continental C975 radial of the M3 Lee. Armored doors covered the direct-vision slits in the hull casting ahead of the driver and co-driver. Tanks of this configuration first saw combat with British forces at al-'Alamayn ("El Alamein") in October of 1942. They also equipped the 2nd Armored Division in North Africa.

Before long, however, the M4A1 was being built with new components. The first change was the elimination of the direct-vision doors from the hull casting. These openings had proven to be weak points in the frontal armor. Next was the introduction of the heavy-duty vertical volute suspension system with the return roller moved to a bracket at the rear of the bogie. This arrangement allowed more space within the bogie casting for larger springs which were better able to cope with the weight of the tank.

Minor changes also made their way into production tanks. All tanks began to leave the factories equipped with sand shields. A blade sight was added to the turret roof ahead of the commander to allow him to traverse the turret quickly to lay the gun on a target.

Other production changes are less obvious in period photographs. The outer shield of the M34 gun mount received ears on the casting that protected the recoil area of the gun barrel from small arms fire. A small shield was added to the coaxial machine gun barrel to keep small arms fire from entering the turret interior. A single-piece cast cover for the differential and final drive assembly was produced and used concurrently with the three-piece design.

Changes were made to the tracks. New designs incorporated cleats or chevron patterns to increase traction. To save scarce rubber, several track designs of all-steel construction were developed. Extended track end connectors known as "duck bills" were developed to decrease the ground pressure and improve flotation on soft ground.

The M34 gun mount was replaced by the new M34A1, with a wider shield that included a direct-vision telescopic gun sight and protected the opening for the coaxial machine gun.

Attempts were made to increase the armored protection of the tank. Appliqué armor plates were added over areas of the hull sponsons to protect internal ammunition stowage. The thin area of turret armor ahead of the gunner received a plate, and then in later models the area was thickened in the original casting. The time of the addition of some of these changes were different at different plants, and some where installed at depot level. This made some odd combinations of features possible.

A total of 6,281 M4A1s were built with the 75mm gun before production ended in December of 1943. They served with all U.S. armored units, including with Army and USMC in the Pacific. The British received 942 M4A1s through Lend-Lease.

M4A1 Small Hatch

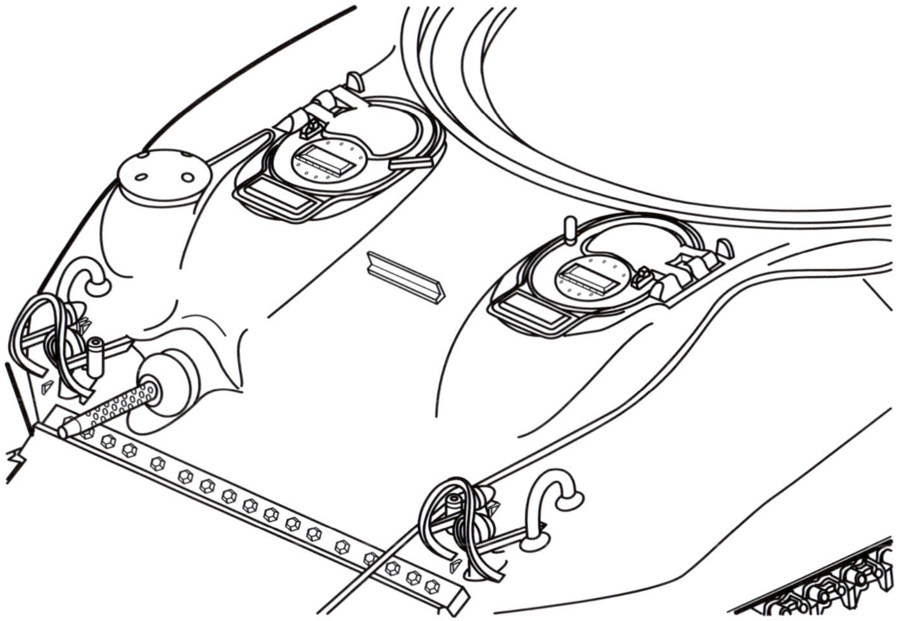

The first production M4A1s had many distinctive features. The turrets had a rotating shield over the gunner's sight in the right side roof. The front hull housed two fixed machine guns to the left of the ball mount above the transmission. With rare exceptions, these early tanks do not appear to have seen combat in U.S. service, remaining stateside instead for training purposes. (National Archives)

The cast-hull M4A1 was the first Sherman produced at Lima Locomotive Works in February 1942. These early-production tanks, undergoing final production at the Lima plant, are equipped with the openings for the fixed box machine guns and armored rotors over the gunner's sight that were soon eliminated from production. (Patton Museum)

Replacement tanks were soon arriving in North Africa with later-production features. This M4A1 at a depot in Wahrân ("Oran"), Algeria, in April 1943, is equipped with the heavy-duty Vertical Volute Suspension System (VVSS) with the return roller trailing the bogie – an arrangement that allowed more room in the bogie housing for larger volute springs. Adorning the hull side are signatures of personnel who have verified a check list before shipping. (National Archives)

This 1st Armored Division M4A1 command tank in Tunisia in February 1943 is commanded by and named for Major Jim Simmerman. The tank has an aerial for a second radio mounted in the opening for the front hull ventilator and its suspension is equipped with the early type of asymmetrical track skid. The geometric camouflage is applied with a coating of local mud. The device on the turret roof is the windshield of the driver's foul-weather hood which was equipped with a defroster and windshield wiper, normally stowed in the lower hull. (Patton Museum)

This early-production M4A1 with the distinctive yellow turret markings of the 1st Armored Division is operating near the al-Qaşrayn ("Kasserine") pass in western Tunisia in February 1943. The tank has a full set of the sand shields that were designed to reduce the dust churned up while moving in dry areas. Many U.S. units quickly removed these shields since they hindered access to the suspension and tracks for maintenance. (Patton Museum)

An M4A1 of 1st Armored Division passes a shepherd and his herd near Lucca, Italy. This is a mid-production tank showing many early features including the M34 gun mount, no external locks on the cupola hatches, and a fender-mounted siren without a protective guard. The tracks are the T54E1 steel chevron type. (National Archives)

M4A1 Shermans assemble in Binzart ("Bizerte"), Tunisia, in preparation for the invasion of Sicily in early July 1943. Markings for the Operation Husky landings include the disruptive camouflage painting of Earth Yellow over the standard Olive Drab and the addition of large wide surrounds around the white stars. A Higgins PT boat undergoes repairs in the background. (National Archives)

This M4A1 with the 752nd Tank Battalion is seen maneuvering through the streets of Pisa, Italy, in August 1944. The 752nd had been activated in 1941 at Ft. Knox, Kentucky, with men from the US West Coast originally filling its ranks. In this view, the co-driver is brandishing a .45-caliber M1928A1 Thompson submachine gun as the tank commander mans the M2 HB .50-caliber antiaircraft machine gun. On the right front fender is lashed a wicker-wrapped bottle, presumably filled with Italian wine.

An M4A1 comes ashore out of the mouth of a Landing Ship, Tank (better known as an LST) during exercises in 1943. No special preparation appears to have been made for deep fording with the exception of plugging the front machine gun mount and removing the headlights. The tank has the single-piece cast E4186 differential cover. The tracks are the T48 rubber chevron style. Extensive training was crucial for the success of amphibious assaults. (National Archives)

An M4A1 of the 760th Tank Battalion advances past knocked-out armor near Maria Infante, Italy, on 14 May 1944. The tank has received a turret appliqué plate and the small rotor shield for the coaxial mounted .30-caliber machine gun. The tracks are the T48 rubber chevron style. (National Archives)

VVSS Evolution

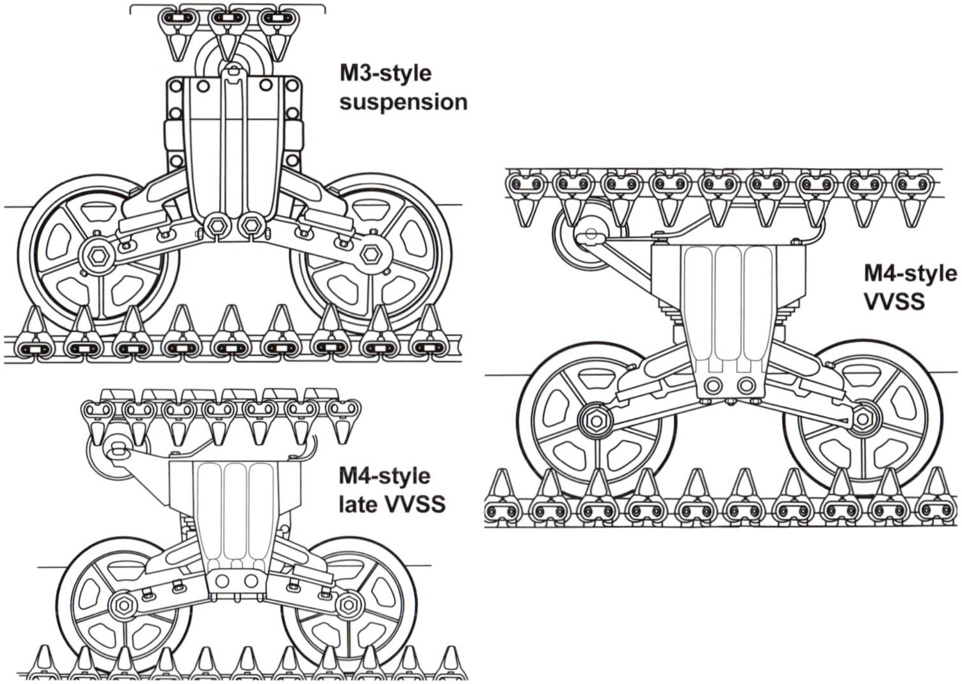

Unit workshops repaired recovered tanks to get them back in combat as quickly as possible. This M4A1 in Italy has been knocked out by a penetration to the right rear of the turret. Tanks worthy of repair were patched and painted; those that could not be repaired were often cannibalized for parts. (Patton Museum)

A pair of M4A1 Shermans knocked out in the Italian countryside. The foreground vehicle has burned and internal ammunition fires have caused explosions which have jettisoned the turret and separated the final drive assembly from the lower hull. (Naval History and Heritage Command)

The rainy season in Italy often brought combat operations to a halt. Armies assumed static positions at times for months. The crew of this 1st Armored Division, 13th Tank Battalion M4A1 has attempted to make their extended stay as dry and comfortable as possible. The hatch openings have been covered with rain slickers and tarpaulins and the tank is littered with the crew's personal gear. Arrayed upon a makeshift table fashioned from pieces of ammunition crates are mess tins, utensils, and a G.I. camp stove. The numerous empty cardboard ration boxes and ammunition packing tubes strewn around the area give some idea as to how long the tank has been sitting in this position. (Patton Museum)

The wrecks of these M4 Sherman tanks were photographed in the area of Cisterna, Italy, in the spring of 1945. Their hulks are testimony to the fighting that raged there a year before. (National Archives)

A rare overhead shot of an M4A1 of the 13th Tank Battalion near Riolo, Italy, in the winter of 1944-1945. American units were prone to carry as much extra stowage as possible. A first-aid kit has been attached to the right turret side. The hoop ahead of the commander's cupola is a guard for an additional sight used for indirect fire roles in the mountainous terrain. (Patton Museum)

Tanks and infantry carry out amphibious maneuvers on a beach of Camp Bradford, Virginia, in 1943, with Navy personnel looking on. The M4A1 is a mid-production variant with T48 tracks, M34 gun mount, heavy-duty suspension, and a three-piece style differential assembly. (Naval History and Heritage Command)

An M4A1 of 10th Armored Division takes part in training maneuvers in 1943. The tank's name, *Duke,* follows the normal practice of using the tank company as the first letter of the tank name. The continued upgrading of Sherman production can be seen from a number of items on this tank, including a mount for a searchlight on the turret, helper springs on the hull hatches, the relocation of the siren to the center front hull behind a guard, and external locks on the commander's hatch plates. (National Archives)

Villagers look on as an M4A1 races through the streets of Aubencheul-au-Bac, France, on 2 September 1944. The tank is very typical of those seen during the battle of Normandy and the breakout from the beachhead. It is equipped with a Culin hedgerow cutter, wide M34A1 gun mount, and a full set of appliqué armor. All of the national star markings have been overpainted. (National Archives)

Another rare type of cast-hull 75mm Sherman was the DD or Duplex Drive tank. These tanks were used during the Normandy landings and had a collapsible canvas flotation screen and two propellers at the rear to enable the tanks to swim. A large percentage of the vehicles were lost in the rough seas on 6 June 1944. After the landings many DD crews removed the bulky canvas structures and fought as conventional armor, like this example. (U.S. Army Engineer School History Office)

The drive across France in 1944 would have been impossible without Army Engineer units bridging the many rivers and streams. Here a 3rd Armored Division M4A1 crosses the Seine via a pontoon bridge at Seine-Port on 26 August. The tank is equipped with the hedgerow cutter and additional sandbag armor. Handles have been added to the turret sides to allow the crew to store additional personal items. (National Archives)

M4A1 Rear Deck

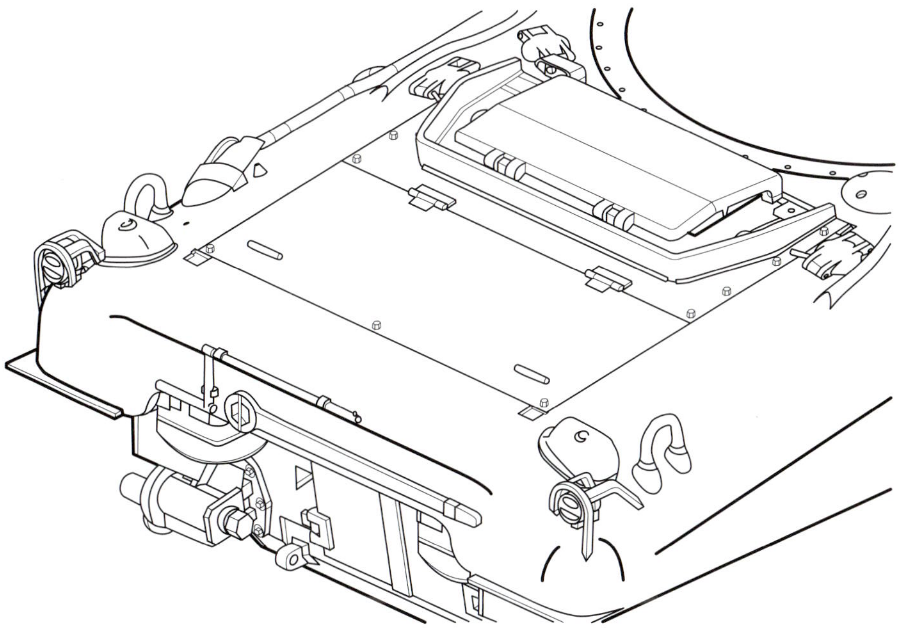

The split hatch cupola found on early-production tanks lacked external locks on the hatch plates and opened to a shallower angle than later hatches. The .50-caliber Browning machine gun has all the features of weapons issued prior to early 1944, with the complex two-position pintle mount and clip for installing the small 50-round ammo box. Maj. Gen. Lindsay Silvester, pictured, was then commander of the 7th Armored Division. (National Archives)

75mm Gun Rotor Shields

M34 early

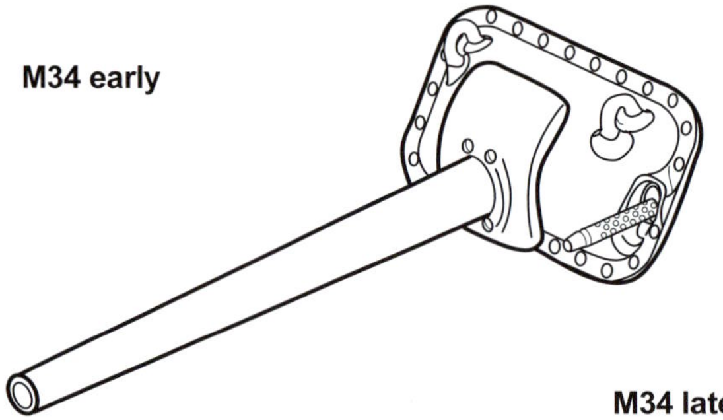

M34 late

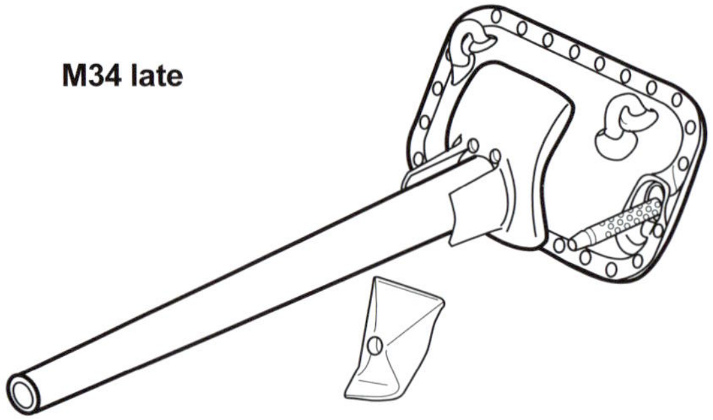

M34A1

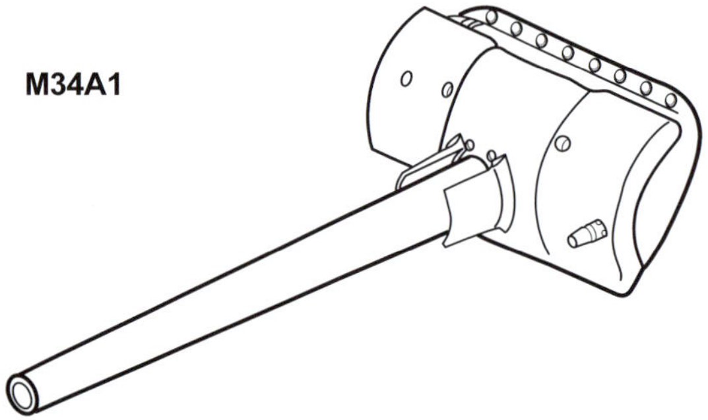

Engine Data

Engine Make/Model	Continental R975-C3	Ford GAA	GM 6046	Chrysler A57
No. of Cylinders	9 radial	60 degree V-8	2 X 6	30
Cubic In. Displacement	973	1,100	850	1,253
Horsepower	400 @ 2,400 RPM	450 @ 2,600 RPM	410 @ 2,100 RPM	425 @ 2,400 RPM
Torque	890 @ 1,800 RPM	950 @ 2,100 RPM	885 @ 1,900 RPM	1,060 @ 1,800 RPM

Many Sherman tanks that had been used in training were sent back through factories for upgrades and overhaul as part of a continued re-manufacturing program. This tank photographed in January of 1945 has many upgrades including an alidad sight, gun travel lock, relocated siren and guard, and short fenders that allow the use of the extended end connectors (not present) with sand shields. (Patton Museum)

A rear view of the same tank shows additional upgrades. Storage brackets for the .50-caliber AA machine gun have been added to the turret rear, and a stowage rack has been added to the rear of the upper hull. Note the plates added to move the sand shields outward to clear duck-bill extended end connectors, and the plugged pistol port door on the turret side. (Patton Museum)

Remanufactured

As tank crews were trained in the United States they were shipped overseas with new tanks. A program was implemented to rebuild the numerous tanks that had been used in the training units. Several factories and depots overhauled and upgraded thousands of early-production Shermans. Chrysler's Detroit Tank Arsenal began the program late in 1943, and rebuilt 1,610 M4A4 tanks by October 1944. The tanks received new components to replace worn out parts and as many upgrade parts as possible were used to bring the vehicles up to the latest standard. These upgraded components included items such as the M34A1 gun mount, gun travel locks, appliqué armor plates, and extended end connectors on the tracks. The tanks received a narrow fender that allowed the sand shields to be installed over the end connectors. The re-manufacturing program differed from programs in which particular items like appliqué plates were added at depot and unit level. The re-manufactured vehicles received a complete mechanical rebuild as well. The majority of the M4A4s were sent to the British.

In August of 1944, the M4, M4A1, M4A2, and M4A3 were also involved in the program. The tanks were sent to Fisher Tank Arsenal; to Chrysler's Evansville, Indiana, plant; to International Harvester; and to the Lima, Chester, and Richmond Tank Depots for the work. In total 5,880 M4 series tanks were re-manufactured.

A few re-manufactured dry-stowage M4A3s were sent to Europe for use by the U.S. armored units in the late winter and spring of 1944-1945 to make good tank losses from the Battle of the Bulge. The re-manufacturing program ended in May of 1945.

Low-bustle Turrets

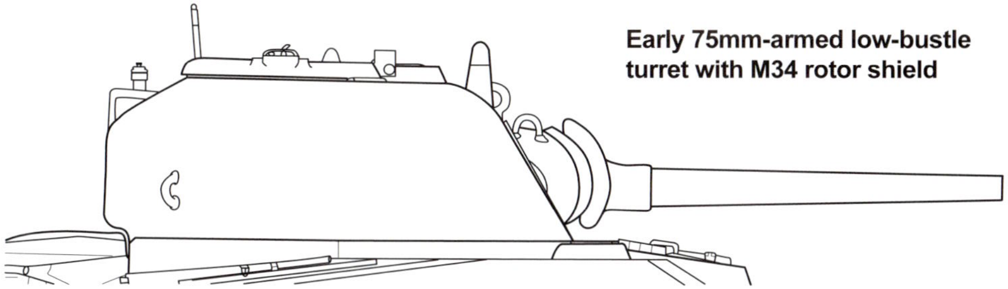

Early 75mm-armed low-bustle turret with M34 rotor shield

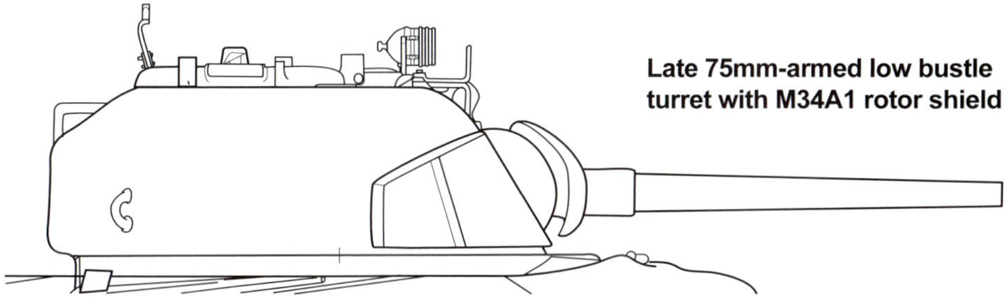

Late 75mm-armed low bustle turret with M34A1 rotor shield

Many of the tanks bound for use against the Japanese were shipped to units through Australia. Here an M4A1 is unloaded at an Australian port in April 1943. Waterproofing has been applied to openings and seams to keep the salt spray from fouling the interior's electrical and mechanical components. All accessory items supplied with the tank are boxed on the rear engine deck. (National Archives)

At the same port in Australia crews set about the task of removing the shipping protection and preparing the tanks for use. The main guns have a brace of wood to keep the gun stationary during shipping, in order to protect the gyro-stabilizer. The second tank in line is still equipped with the narrow M34 gun mount. (National Archives)

A Marine Corps M4A1 in Australia trains for combat. The crew has removed the headlights and stored them in the interior and installed the foul-weather cover on the front machine gun. The tank is equipped with the M34A1 gun mount, and T49 three-bar cleat steel tracks. (National Archives)

These Marine Corps M4A1s are all from similar production batches as those seen in Australia. The tanks were attached to the 1st Marine Division and are shown here along a beach in the aftermath of the Arawe battles on 27 March 1944. The tanks are equipped with the T49 three-bar cleat style track. To the left is the bow ramp of an LCVP landing craft. (Patton Museum)

Shanghai Lil, an Army M4A1, carries infantry inland from the beach. The tank is a part of the landings at Humboldt Bay in southern New Guinea, 22 April 1944. These landings were the beginning of MacArthur's push toward the Philippines. (Patton Museum)

This column of Shermans pauses on Wake Island on 1 August 1944, just days after the liberation of the island had been completed. These tanks are equipped with the T49 cast steel block-type track. (Patton Museum)

An M4A1 drives down a road near Mokmer Strip on Biak, New Guinea. In May 1944, the U.S. Army expended considerable effort to capture this airfield, located less than 100 yards from the beach. Continued Japanese presence in the high ground above the airfield delayed U.S. use of the base for aerial operations, however.

Narrow Track Types

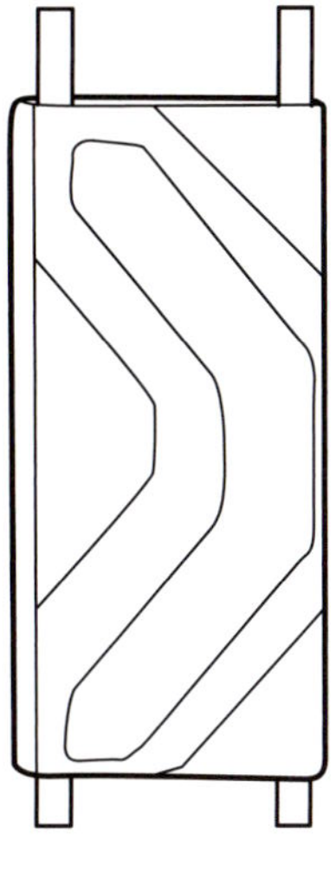

T48 Chevron

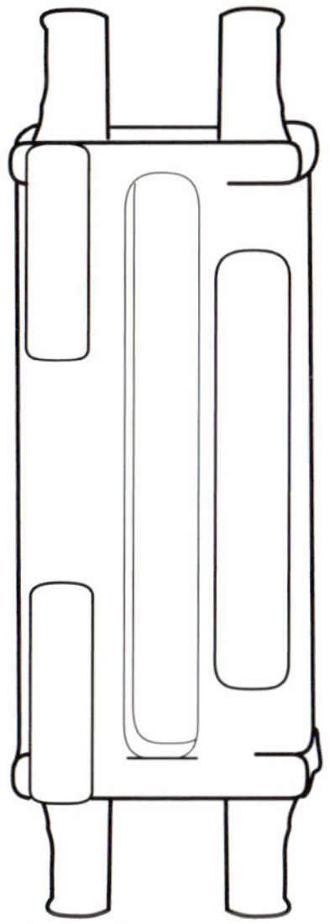

T49 Cast Block

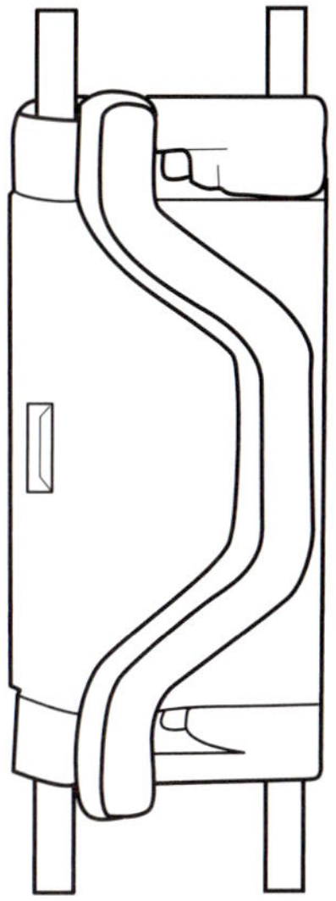

T54E1 Truscon

Small-Hatch M4A2

The first M4 series welded-hull tank to be built was the M4A2. Production began at both the Fisher Tank Arsenal, a division of General Motors, and the Pullman Standard Car Company in April of 1942. The M4A2 was powered by the General Motors 6046 diesel engine, which was produced by coupling two six-cylinder truck engines to a single fluid-driven torque converter to transmit power through a single drive shaft. The maximum output of the engine was 410 bhp, and performance during trials was superior to the radial-engine powered tanks. There was some thought of outfitting the Army with the diesel-powered tank, but there was resistance to supplying diesel fuel for tanks and gasoline for other vehicles. The Navy used diesel in smaller craft, so a supply of fuel was available for the M4A2s used by the Marine Corps.

The initial batch of tanks was built with direct-vision slots protected by a hinged armored door ahead of the driver and co-driver positions. The turrets, suspensions, and differential covers were identical to the types used on the M4 and M4A1. The full range of development changes were incorporated through production of the M4A2.

During 1943 Fisher tank arsenal constructed a series of unique M4A2s on which all of the cast components of the hull were replaced with pieces made of rolled steel plate. These items included the driver's hoods, splash rail around the turret, antenna socket, and front machine gun mount. The hard edges and rectangular shape of the driver's hood front plates give these tanks a distinctive appearance. Many of the small-hatch M4A2 tanks used by the Marine Corps were of this type.

The M4A2 was the first welded hull Sherman to be produced. The new tank carried over the existing suspension from the M3 Medium tanks with the return roller centered above the bogie assembly. (National Archives)

M4A2 Small Hatch with Welded Hoods

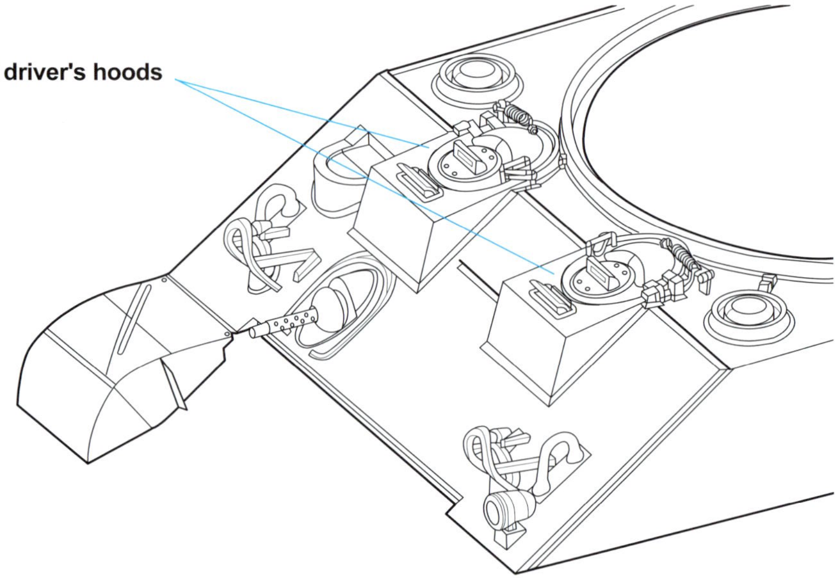

The M4A2 was powered by the General Motors 6046 diesel engine. The power plant was derived by coupling two 6-cylinder G.M. bus engines to a single torque converter unit that ran a single central drive shaft. This allowed the twin engines to be adapted to the standard Sherman layout and transmission. (National Archives)

As production of the M4A2 continued, many later features were incorporated. This M4A2 has a fairly common combination of details for an early-production tank: the single-piece cast E4186 differential cover, heavy-duty VVSS bogies, direct-vision doors on the driver's hoods, and the narrow M34 gun mount with protective ears on the outer rotor shield. (Patton Museum)

While normally a Lend-Lease or Marine Corps vehicle, the M4A2 served the U.S. Army as an important training vehicle within the U.S. Here M4A2s of the Armored Force School maneuver with an M3A1 Medium tank in the background. The markings at this time could be white or yellow, with the AFS in black on the turret star, and registration numbers in blue drab. The vehicle in the foreground carries the name *Beelzebub* on the side of its hull. (Patton Museum)

The Sherman first saw combat in the Pacific when the Marines took it ashore on the Tarawa Atoll in the Gilbert Islands from 20-23 November 1943. C Company, 1 Marine Amphibious Tank Battalion attempted to land 14 M4A2s, but most were knocked out or swamped in shell holes before making it ashore. This after-action shot of *Charlie* reveals just how fierce the fighting was on the tiny island. (Patton Museum)

Colorado is a "C" Company tank that landed on Tarawa after the fighting. These tanks shared many common features, including the welded driver's hoods, T54E1 tracks, M34A1 wide gun mounts, and a rack on the rear hull for Jerry cans filled with water. Emblazoned in yellow on the forward portion of the sides of the hull were the name of the tank and a trumpeting elephant emblem. The cleats welded to the hull sides were for stringing barbed wire. (Patton Museum)

This Marine M4A2 on Saipan in June or July of 1944 carries no visible markings other than a company and vehicle number on the turret side. This shot provides an excellent view of the details of the welded driver's hoods, front machine gun ball mount, antenna mount, and splash rails of Fisher-built tanks of this period. The crewmen wear the Tank Helmet and are down to shirtsleeve order in the tropical heat. (Patton Museum)

The Japanese had no tanks to match the Sherman in combat, and U.S. units seldom faced massed Japanese armor. This shot of an M4A2 damaged on Tarawa demonstrates, however, that the 47mm anti-tank gun was capable of penetrating the side armor of the Sherman. (Patton Museum)

A 2nd Marine Division M4A2 supported by infantry advances along a road on Saipan, on 7 July 1944. The distinctive shape of the driver's hoods identifies this as a welded-hood M4A2 manufactured by the General Motors Fisher Tank Arsenal. The gun mount is the wide M34A1 type, and the tracks are T54E1 steel-chevron style. All Marine small-hatch M4A2 tanks appear to have been of this type. (Patton Museum)

A 4th Marine Division M4A2 named *Corsair* comes ashore with LVTs on Tinian, on 25 July 1944. The tank is equipped with deep wading stacks, a precaution against the tank's engine and interior being swamped while in the surf. The liquid containers on the left rear corner carry water rather than fuel. In the heat of the South Pacific, tanks and Marines required large amounts of fresh water. (Patton Museum)

Colorado is one of four M4A2 tanks assigned to C Company, 1st Corps Tank Battalion (medium), 3rd Platoon that made it onto Red Beach 3 on Tarawa. She then became the unit's only tank to survive the first day. The tank took two hits, the second of which set the vehicle ablaze. *Colorado's* crew put out the fire by driving into the surf.

Marines examine a mid-production M4A2 of the 3rd Tank Battalion. The side hull plate appears to be a patch of an earlier hull penetration. Forward of the factory-style exhaust stack is a pile of the crew's personal gear atop the tank's engine deck.

Ateball, an M4A2 assigned to A Company, 3rd Tank Battalion on Iwo Jima, sports appliqué armor plates that were obviously field installed. Just visible on the rear deck is the adapter for a deep-water air-intake system.

Three mid-production M4A2s from the USMC 3rd Tank Battalion roll along a trail. There are field-applied side hull armor plates, the rear hull top rebar rail, and the markings that include a star and bar on the hull side, numbers on the rotor shield and "U.S. Marines" and eagle, globe, and anchor emblem on the transmission housing.

A 3rd Marine Tank Battalion mid-production M4A2 advances on one of Iwo Jima's few roads during February of 1945. By this late date this tank is an old war veteran, built in the period before factory appliqué plates were installed. The left hull side plate is locally made and is marked with the name *Atlae.* (National Archives)

Small-Hatch M4A3 75mm

With Continental radial engines in short supply, the armored board began looking for alternative tank power plants that could be produced in existing facilities. Ford's relatively compact GAA Petrol V-8 engine was a modification of an experimental 500-h.p. V-12 aircraft engine. It fit quite well into the Sherman's engine compartment, which had been designed for the tall radial engines. Its drive shaft ran low below the turret basket and connected to the existing Sherman transmission. Subsequent modifications to install the Ford engine brought about the M4A3. By the time production ceased in September 1943, 1,690 M4A3 dry-stowage 75mm tanks had been completed.

The first M4A3s to see combat were later-production M4A3 wet-stowage tanks that began to arrive in Europe in the early fall of 1944. Output of the M4A3 (75)W reached 3,071 between February 1944 and March 1945. The tanks featured the simplified hull construction with the 47° single-piece 2½-inch thick glacis armor plate. The hull hatches were a new larger oval counterbalanced design in a cast plate of the hull roof. These hatches were far easier to enter and exit, an important asset for crews escaping a burning vehicle. The bulk of the new tanks were built with the high bustle turret with a hatch in the roof for the loader, and the M34A1 mount for the M3 75mm gun. The differential covers, which can be identified by their more pointed cross section, were all of the late E8543 design, which had thicker armor, improved brakes, and dual-ear tow lugs.

Initially, all M4A3 dry stowage tanks remained in the U.S. for training. The tanks in England slated for the invasion of France were all radial-engine M4s and M4A1s. Introducing the new tank would have complicated the supply of spare parts to the theatre. After the German Ardennes offensive, however, a few remanufactured dry-stowage M4A3s were shipped to Europe to make up the resultant tank shortage.

U.S. forces preferred the M4A3 type. Later variants were the basis of 76mm-gunned tanks, 105 howitzer tanks, and many received the later wide-tracked HVSS suspensions.

This dry-stowage small-hatch M4A3 has a low-bustle turret with M34 gun mount. These dry-stowage tanks remained stateside for training during 1944 to avoid overburdening the supply line with spare parts for a new tank engine. (National Archives)

An M4A3 of the 5th Armored Division bursts through the pastoral fall foliage during training in the United States. The tank is equipped with the single-piece cast E4186 differential cover. There are fixed periscopes in the driver's hood, the earlier direct-vision doors having been eliminated by this time. The tracks are the T49 steel cleat style. (National Archives)

Many of the dry-stowage M4A3s that had served as training tanks in the U.S. were sent to Europe to make up for losses during the Battle of the Bulge. These tanks had gone through a re-manufacturing program and were updated with many late features. This M4A3 can be identified as an earlier variant by the appliqué plate on the left hull side and ventilator dome visible by the driver's hatch. A narrow fender has been installed to cover the extended track end connectors. (Patton Museum)

The quest for suitable tank power plants brought about the A57 Multibank, a complex system gearing a group of five Chrysler six-cylinder automobile engines around a common output shaft. To accept the larger engine, the lower hull of the M4A4 was lengthened by 11 inches in comparison to other Sherman tanks. (National Archives)

This early-production direct-vision M4A4 was photographed while undergoing tests at Aberdeen Proving Grounds in May 1942. M4A4s entire production received the three-piece differential/final drive assemblies. The bogies have been re-spaced for the longer, lower hull, and the T48 rubber chevron track is installed. (National Archives)

Small-Hatch M4A4

One of the key criteria in tank design has been a lightweight power plant with a high horsepower-to-weight ratio. With such considerations on their minds, engineers in many instances turned to aircraft power plants to drive their tanks. With demand from the front line rising, it is not surprising that the aircraft engine industry was already straining to meet output quotas, and accordingly the search was on for alternative power plants for tanks.

One of the alternative power plants was the Chrysler A57 "multibank" engine. This unique power plant was in essence five six-cylinder truck engines connected to a common crankcase. The hull of the M4A4 had to be lengthened 11 inches to accommodate the engine. This change, in turn, required a difference in the spacing of the bogies on the side of the hull and an increase in the number of track blocks from 79 to 83 per side. The bogie spacing is a distinguishing trait in photographs. All M4A4s were built with the three-piece differential/final drive housing.

Chrysler manufactured the first of the M4A4 tanks in June 1942, and by the time it had finished work on the last of the design, in September 1943, 7,499 of the unorthodox machines had been produced. Most of these tanks were supplied to the British under Lend-Lease, although they were used for training U.S. units inside the United States as well. Chrysler re-manufactured 1,610 of these training tanks between December 1943 and October 1944, and upon completion, all of those vehicles were shipped out to the British.

All of the M4A4 tanks were armed with the 75mm gun. Initial production batches were equipped with direct-vision doors on the driver's hoods. The early M4A4 turrets carried the M34 main gun mount, while the later ones were fitted with the M34A1 mount.

M4A4 Rear Deck

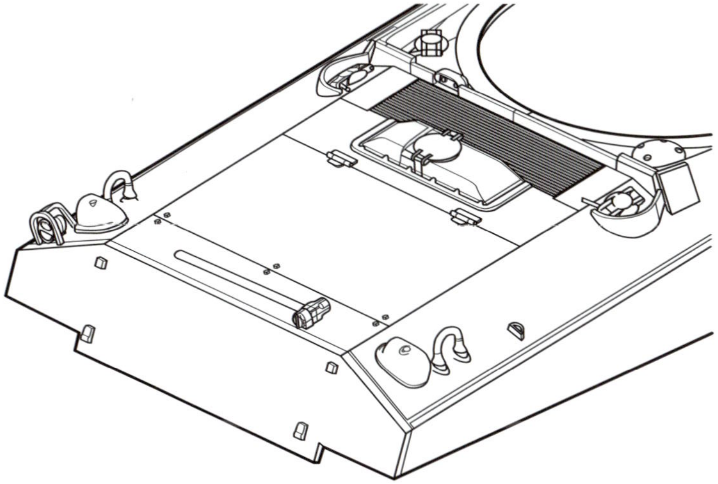

The crew performs maintenance on this M4A4 at the Desert Training Center in California. The tank displays many early features, including the M34 gun mount with narrow rotor shield, direct-vision doors, two-position pintle mount for the .50-caliber Browning, and a lack of appliqué armor. The half-track is equipped with an air compressor which is pressurizing the grease line extended to the tank. (National Archives)

M4A4 tanks assemble for maintenance at the Desert Training Center in California. The tanks have both direct-vision and fixed-periscope style driver's hoods. The M34 mounts have the narrow rotor shields lacking the protective ears. The tracks are a mixture of steel and rubber chevron designs. (National Archives)

While U.S. units did not use the M4A4 in combat, the tank was widely used in stateside training. Here an M4A4 operates along the Alaska Highway for winter training during February of 1944. The exhaust deflector is seldom seen in this lowered position, and was often removed by crews. The commander's hatch is covered by a clear plastic cylinder to allow a full range of vision while keeping heat in. (TACOM LCMC History Office)

Re-manufactured M4A4s are unloaded from rail cars of the Bengal Assam Railway at Ledo, Assam, India. At the foot of the Himalayas, Ledo marks the beginning of the Stillwell Road. It was common for tanks used for stateside training to return to the factory for overhaul and upgrade before being shipped to Lend-Lease users. These tanks are headed for use in Burma ("Myanmar"). (U.S. Army Transportation Museum)

An M4A4 with direct-vision doors undergoes maintenance and repairs during training maneuvers in the desert. The tank belongs to 7th Armored Division. The white stripe around the turret and white stars are common on tanks in training in both the United States and England. The bogies have an early symmetrical track skid. (National Archives)

Lucky Eleven is an M4A4 of the Chinese Provisional Tank Battalion operating in Burma, on 25 February 1945. These tanks were crewed by a mix of Chinese and Americans and formed the bulk of the tank strength along the Burma Road. An additional name, *Martha Sue,* is visible on the turret side near the smoke mortar, as well as a British census number on the hull side. Many of these tanks later fought in the hands of Nationalist Forces during the Chinese Civil War. (National Archives)

M4 Composite

The final production variant of the M4 75mm tank incorporated a cast section for the front armor welded to an assembly of rolled plates in the rear. The tanks have become known as "composite" or "hybrid" hulls. The cast section extended from the differential/final drive housing back to the splash rail forward of the turret, and had openings for the front machine gun and driver's hatches. Production began at the Chrysler Tank Arsenal in August of 1943, and continued through the end of 75mm M4 tank production in January 1944. The initial vehicles built were a small hatch variety, and can be identified by the domed ventilator fans outboard of the hatches in the forward casting. There were perhaps fewer than 100 of these early models built. The design then switched to the newer larger counterbalanced hatches with a blower mounted centrally in the hull roof.

The tanks retained the dry stowage racks for ammunition, and were thus equipped with appliqué armor on the hull sides at the factory. M4 components were going through a bit of a transition at the time of construction of the composite hull tanks. They can be seen with a mix of late low bustle turrets that lack the pistol port on the left side and have the increased armor thickness in front of the gunner. Other later models have high bustle turrets with the oval loader's hatch in the roof and the pistol port reintroduced.

The return roller brackets are the straight style initially, later the upswept style are used. All of the tanks receive the late-style E8543 differential/final drive housing. The tanks were powered by the R975-C1 Continental radial air cooled engine.

Approximately 1,675 of the composite-hull tanks were built. Numbers in specific units are difficult to trace as units made no differentiation between them and earlier all-welded M4s. The tanks seem to be relatively rare with U.S. units in Europe, but a number of them were sent to Army tank units in the Pacific, and to the British through Lend-Lease. The British converted some of their M4 composite-hull tanks to Fireflies with the potent 17-pounder gun.

Workers at the Chrysler Tank Arsenal in Detroit prepare the first M4 composite-hull tank for rail shipment. Composite-hull production began with the very rare (perhaps fewer than 100 built) small-hatch hull. These can be identified by the bulges for ventilator fans outboard of the hatches in the hull roof. (National Archives)

This 6th Armored Division M4 composite is seen knocked out near Longvilly, Belgium, on 16 January 1945. The Division's 15th Tank Battalion took part in efforts to break out of the Bastogne area following the German offensive of December. The tank displays the large tactical number on the hull side in white and the Wolf Head emblem over the armored tri-color triangle on a white square. (Patton Museum)

M4 Composite Hull Front

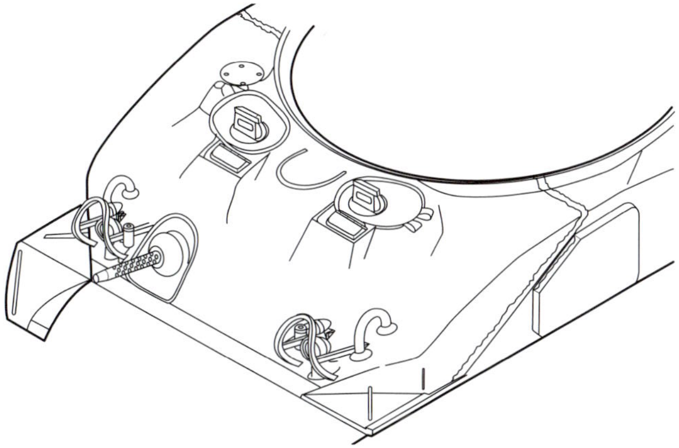

A group of M4 and M4 composites train in the English countryside prior to being shipped to France. The tanks carry a complete set of markings including bridging classification on the hull front of the foreground vehicle. M4 composites were mechanically identical to the M4 and M4A1 and were often mixed into units equipped with these tanks. (Patton Museum)

An Army M4 composite advances in support of Marine Infantry in September 1944. The composite-hull Shermans were built at a time of transition in production and can have various types of suspension and turrets. This tank displays many later features such as a high bustle turret, upswept brackets for the return rollers, and the stowage rack on the rear of the hull. (National Archives)

Broadway Terror, a composite-hull M4 of the 44th Tank Battalion, took part in a lightning U.S. assault that captured the internment camp at Santo Tomas University in Manila in February 1945. The civilians inspecting the tanks are internees freed by the stunning U.S. attack that foiled Japanese plans to liquidate them. The geometric markings on the hull side are unique to the unit, and their exact meaning is unknown. (National Archives)

The contours of the front cast portion of the M4 composite hull appear in excellent detail in this photograph of an Army M4 taking part in the Peleliu battle of September 1944. It is possible that the tank belongs to the 763rd Tank Battalion, which is known to have been equipped with this type, and supported the 81st Division at this time. (Naval History and Heritage Command)

An M4 composite hull overlooks Baguio, Luzon, from a nearby hillside on 27 April 1945. The tank carries a number of interesting markings, which include two names. *Big Mike* is painted on the gun barrel, while *Brassierless Baby* is on the left hull side, and a skull face with the number 4 in the forehead just to the rear. A length of spare track is welded to the turret side for additional protection. (Patton Museum)

M4s were converted to main armament flamethrower tanks prior to the Okinawa landings. 54 tanks were assigned to the Army's 713th Tank Battalion to support all divisions during the fighting. This M4 composite is engaged in support of 6th Marine Division infantry on 19 June 1945. An interesting detail is the late vision cupola on the low bustle turret casting. (Patton Museum)

Another of the main armament flamethrower tanks on Okinawa fires a flame burst on 25 June 1945. The flame tanks maintained a good reputation for protecting the crew and safe storage of the flame propellant. The stars on the turret and hull side have been painted out with black or fresh Olive Drab paint. The tank is fitted with "duck bill" extended end connectors, a rear stowage rack, and vision cupola. (National Archives)

M4 105mm Howitzer Tank

Part of the original concept of the M4 Sherman design was to use the 105mm howitzer as alternative armament. Development of the 105mm-armed variant began shortly after production started in 1942. The weapon developed was standardized as the 105mm Howitzer M4 and the Combination Mount M52. Although the mount could be used in the standard 75mm turret in production the gyro-stabilizer and power traverse were eliminated, and only a partial turret basket was employed. The new tanks featured the new high-bustle turret castings with a pistol port in the left side wall and a hatch for the loader in the left turret roof. A second ventilator fan was added to the turret rear roof above the radio.

All of the 105mm howitzer tanks were built at the Chrysler tank arsenal in Detroit. From February 1944 until the end of production in March 1945, 1,641 M4 (105)s were built, the final 841 being equipped with HVSS suspension and wide tracks. Originally fitted with the standard split-hatch commander's hatch, the vision cupola was introduced to production once sufficient supply was available around June of 1944.

All of the production of the late simplified welded-hull M4 with single-piece glacis and large counterbalanced hatches received the 105mm howitzer. Ammunition stowage was moved to 66-round dry stowage racks on the hull floor. Fitted between the air cleaners on the rear of the hull was an exhaust-deflecting grille to keep exhaust and cooling air from churning up dust and debris behind the vehicle. The suspensions on the Vertical Volute Suspension System (VVSS) tanks had the upswept bracket for the return roller wheels.

These tanks were assigned to the headquarters companies of armored units in Europe. They began to arrive in combat units shortly after the Normandy landings.

The M4 (105) was powered by the air cooled Continental radial R975 C1 gasoline engine. With the U.S. Army's desire to standardize its combat units with the Ford V-8 powered M4A3, the later-production M4 (105)s were supplied to Commonwealth Armies through the Lend-Lease program.

Large-hatch Configuration

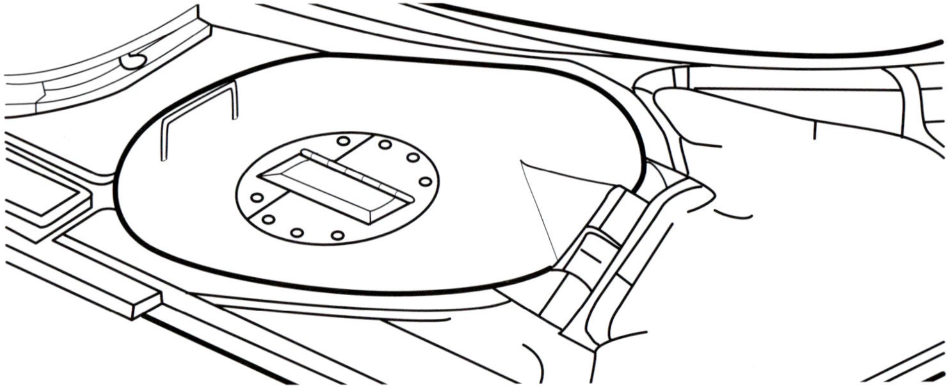

This factory-fresh M4 (105) howitzer tank, photographed on 9 May 1944, exhibits typical characteristics of that time. All late welded-hull M4 production tanks received the 105mm howitzer. The turret is equipped with the standard split-hatch cupola for the commander. Later variants would receive vision cupolas and a special machine-gun pintle mount above the turret's rear ventilator.

Identifying features of the M4 (105) seen from overhead include the second blower dome at the rear of the turret, the armored cover for the engine air intake, and the bracket for stowing gun cleaning rods and aiming stakes on the left rear hull. The additional armored filler cover for lubricating oil on the rear engine deck is a common feature on late-production radial-engine Shermans. (TACOM LCMC History Office)

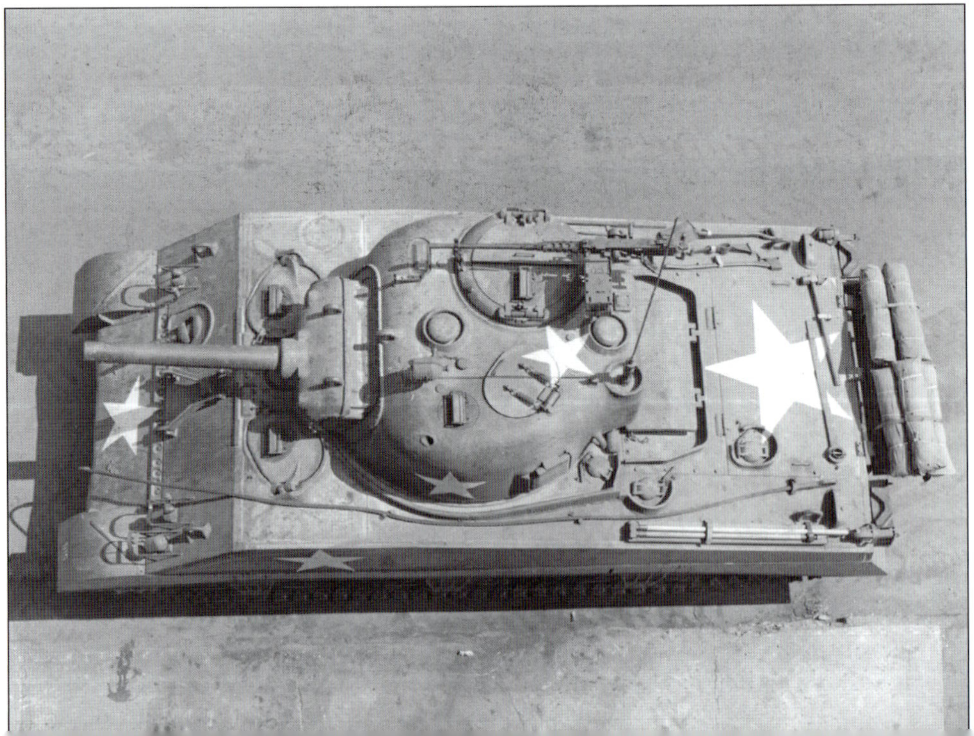

A 6th Armored Division M4 (105) Howitzer tank advances on a country lane in France in the late summer of 1944. The crew has made attempts to obscure the white star on the differential cover. The other markings are intact on the front hull, including the yellow bridge weight classification marking. Both headlights have blackout covers installed. (Patton Museum)

Two M4 (105) howitzer tanks of the 750th tank Battalion fire from fixed positions near Manhay, Belgium, on 30 December 1944. Nets have been set up to camouflage the emplacement. A bugle is among the personal items of the crew that litter the front plate of the vehicle in the foreground. The tank is equipped with T48 rubber chevron tracks. (National Archives)

The crew of a 6th Armored Division, 15th Tank Battalion M4 (105) warm themselves beside their tank in Luxembourg in January 1945. Fiber ammunition packing tubes have been used to form a windbreak around their fire. The large tactical number has been left clean of whitewash, and the area over the unit's wolf head insignia has been cleaned. The howitzer tanks had quite an appetite for ammunition and crates and packing tubes are usually nearby. (National Archives)

The crew of an M4 (105) of an unknown unit open Christmas parcels on 30 December 1944. A bivouac has been set up along the tank's hull side to provide some protection from the snow and cold. A board between the front fenders forms a shelf for stowage. The tank tows an M10 Armored Ammunition Trailer. (National Archives)

The guns on this group of M4 (105) howitzer tanks are elevated near the maximum. Logs and evergreen branches have been positioned on the tanks to break up the silhouette of each vehicle. There are large amounts of stowage on the rear decks of the tanks, and bars have been welded to the turret sides to stow additional gear. An M3 half track is just visible behind the foreground vehicle. (Patton Museum)

An M4 (105) of Headquarters Company 38th Armored Infantry, 7th Armored Division, fires across the Rhine on 13 March 1945. The sides of the turret and hull have been covered with Sommerfield matting, a type of metal screen, to attach camouflage to the tank. All markings on the hull and turret side have been painted out. Ammunition has been laid out atop packing tubes to keep the rounds out of the mud. (National Archives)

Most late-production M4 (105) howitzer tanks were slated for Lend-Lease. The U.S. Army had attempted to standardize on the Ford GAA V-8 of the M4A3 for its use. The tank shortages created by losses during the Battle of the Bulge forced the Army to supply some units with tanks equipped with other power plants. *Panther,* a late-production tank with a vision cupola, is seen in a mixed group of vehicles of the 20th Armored Division in April of 1945. (National Archives)

The final production batches of M4 (105) tanks received the wide track and horizontal volute spring suspension. Tanks with this suspension received the E8 designation, prompting the nickname "Easy Eight." The entire production of M4E8 (105) tanks were sent to the British under Lend-Lease. (Patton Museum)

The new large-hatch wet-stowage M4A1 with the 76mm gun went into production at Pressed Steel Car Company in January of 1944. The tanks were equipped with the new E8543 differential/final drive cover. This early-production example has the M1A1 76mm gun that has no provision for attaching a muzzle brake. (Patton Museum)

The new M4A1 (76) was introduced to U.S. units in preparation for "Operation Cobra," which was the breakout from the Normandy beachhead in late July 1944. The 76mm gun penetrated armor better than the 75mm did, but was still no match for the frontal armor of the German Panther. Here two early production M4A1 76 (W) tanks have been knocked out in the streets of Schevenhütte, Germany, in September 1944. (Patton Museum)

Large-Hatch M4A1 (76)W

The changes made to the welded hull M4A2 and M4A3 models were incorporated into the M4A1 cast-hull tanks. Pressed Steel Car Company began producing the new M4A1 with its 76mm gun in January 1944. The contours of the front of the hull were changed to incorporate the new, larger, counterbalanced hull hatches for the driver and co-driver. The design of the new 76mm turret derived from that used in the T23 Medium Tank program. The earliest versions were built with split circular hatches for the loader, the smooth-barrel M1A1 76mm gun, and they lacked the armored turret vent cover on the rear of the turret. Later turrets had a cast opening for a simplified oval loader's hatch in the roof. These turrets also had a fixed pintle mount for the anti-aircraft .50-caliber Browning machine gun.

The first production models were equipped with the M1A1 76mm gun with no provision for a muzzle brake. They also featured split-hatch cupolas for the loader and straight return-roller brackets on the VVSS bogies. The ammunition was stored in the new wet stowage racks in the lower hull. Soon the later M1A2 and A3 guns were adopted, and return-roller brackets, upswept to eliminate the need for a spacer, were introduced. The oval loader's hatch turret was introduced on vehicle number 1225 in August 1944.

The 76mm tanks were in production well before the Normandy invasion, but the decision was made not to create more logistical problems by introducing new vehicles, parts, and ammunition into the already taxed supply line. The tanks were first issued to troops in preparation for "Operation Cobra," codename for the breakout operation from the Normandy beachheads, in July of 1944. These vehicles were the first 76mm tanks to see combat and were needed for their greater armor penetration ability when facing the German Panther. They became more common with armored divisions as the European battles continued into the fall and winter of 1944-1945.

The last month's production of M4A1(75)s at Pressed Steel Car featured larger hatches. Many of these vehicles were converted to Duplex-drive amphibious tanks. Some of these tanks appear in period pictures after having the canvas flotation screen removed to allow them to function as a standard gun tank.

The M4A1E8 (76)W marks the final development of the cast-hull M4A1 Sherman variant. Production of the large-hatch hull tanks with the T23 76mm turret began in January of 1944, all of the tanks being manufactured by Pressed Steel Car Company of Hegewisch Station in Chicago, Illinois. The late-production tanks were built already equipped with the horizontal volute suspension system with the 23-inch wide track. Production reached 3,426 units by war's end, and 1,327 are thought to have been built as E8 HVSS models. Some VVSS models may have been converted post war.

There is no photographic evidence that the M4A1E8 tanks made it to combat units in Northwestern Europe before the end of hostilities. While the Army quickly standardized on the M4A3E8 for front-line use after the war, the M4A1E8 populated Army Reserve and National Guard tank units for many years. The cast-hull 76mm tanks also soldiered on in many post-war armies. Decades later, Israeli forces would convert a large number of the cast-hull 76mm tanks to M-51 Ishermans.

A group of vehicles from an unknown unit cross a river via a section of Bailey bridge. The M4A1 (76) has Sommerfield matting tacked to the hull sides and turret for attaching branches for camouflage. There is a .30-caliber Browning machine gun on a non-standard mount ahead of the commander's cupola. Typical of late-war U.S. tanks in the field, the engine deck and turret are covered with stowage and personal gear. The bridging equipment and mixture of vehicles in the area give evidence of the huge amount of logistical support necessary for armored operations. (U.S. Army Transportation Museum)

Soldiers in winter camouflage suits board an M4A1 (76) of the 741st Tank Battalion near Schöneseiffen, Germany, on 2 February 1945. The tank has Sommerfield matting applied for attaching camouflage, and a fresh coat of whitewash. The T48 rubber chevron tracks have "duck bill" end connectors mounted. (National Archives)

The crew of an M4A1 (76) of 3rd Armored Division halt on a road near Korbach, Germany on 30 March 1945. The tank has received two additional layers of armor plate to the hull front. Brackets have been welded to support the armor against the contours of the cast hull. Note that the vision cupola and split loaders hatch positions have been reversed on the turret. (National Archives)

The final production M4A1 tanks were equipped with the HVSS suspension and 23-inch wide track. Late production details include rear view mirrors, M1A2 76mm gun with muzzle brake, and oval loader's hatch replacing the split-hatch cupola. No photographs have come to light of these vehicles in European combat, but they served in U.S. National Guard and Army Reserve units, as well as foreign armies for many years following the war. (Patton Museum)

A mixed column of M4A1 and M4A3 tanks from the 771st Tank Battalion enter Moers, Germany, on 24 March 1945. All the tanks have supplemental sandbag armor on their front plates, and many of the tanks have the extended track end connectors for better cross-country performance. All markings have been overpainted, and the bumper codes have been repainted on the differential covers below their original position, which is covered by the sandbags. These code markings appear in the original print to have been carefully painted by hand rather than with the use of stencils. (National Archives)

Large-Hatch M4A2

But for a small initial production with the low-bustle turrets, the M4A2 75mm tanks with large hull hatches were equipped with high-bustle turrets that featured loader's hatches. A few of the vehicles were used by the Marine Corps. These tanks retained dry ammunition stowage and appliqué plates on the hull, unlike similar M4A3 tanks.

Final-production vehicles received the 76mm turret and some were equipped with the wide 23-inch track and HVSS. These vehicles were shipped to the Soviets, and saw post-war service with the Canadians. About 1,300 M4A2 (76)Ws were made with HVSS.

The M4A2 was issued through the Lend-Lease program to the Commonwealth Armies, the Soviet Union, and Free French. Many of the tanks that were shipped to other armies were refurbished vehicles that had served with U.S. Army training units. Over 8,000 M4A2 75mm and 2,915 M4A2 76mm tanks had been built when production ended in May of 1945.

High-bustle 75mm turret with M34A1

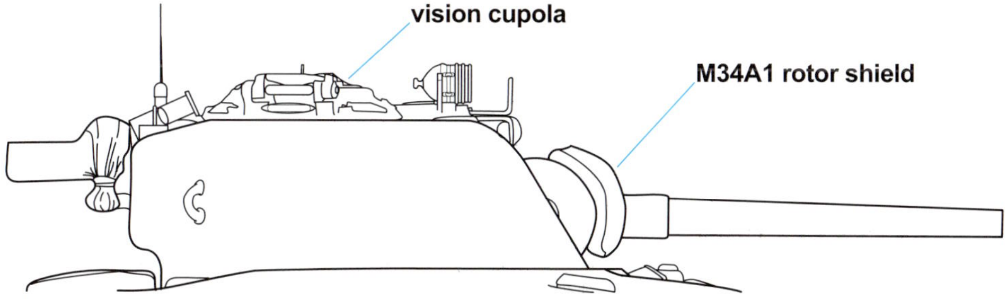

M4A2 Rear Deck Arrangement

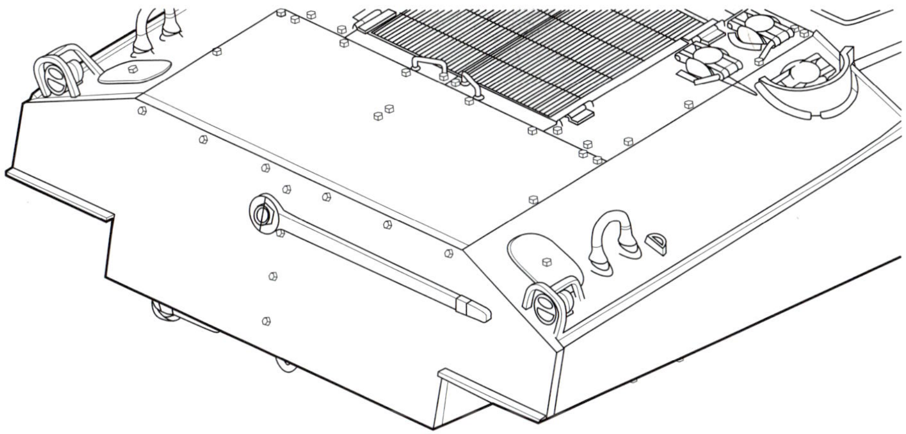

The commander of this Marine Corps M4A2 has dismounted to direct the driver to stop the tank for resupply. This is a late-production tank on Peleliu during September of 1944. The differential cover is the late style with thicker armor and double brakes, identified by the two towing lugs per side. Lengths of spare track have been welded to the hull front and turret sides for additional armor protection. (National Archives)

The final tank combat of WWII involving U.S. forces took place on Okinawa. This 1st Marine Tank Battalion late-production M4A2 is operating with the remnants of its deep wading trunk in April of 1945. The tank has spare track attached to the turret for added protection. The hull hatches have pieces of PSP plating cut to shape and attached for stand-off protection from charges that were carried by Japanese infantry. There are long sections of a woven wicker material attached to the hull sides, which is contrary to the normal procedure of using track blocks or lumber.

A late production M4A2 moves up off the beach into the jungle of Peleliu in September 1944. The tank carries the speed number markings on the turret for quick identification by fellow troops. The tank is equipped with wading stacks and the late high-bustle 75mm turret with an oval roof hatch for the loader. (Naval History and Heritage Command)

Comet, a late-production M4A2, advances with Marine support on Tinian in late July 1944. The 2nd and 4th Marine Divisions that had captured Saipan were immediately landed on Tinian, contrary to the normal Marine Corps practice of allowing a unit a period of recovery after a landing. (Naval History and Heritage Command)

Marine M4A2s of the 1st Tank Battalion capture the airfield on Peleliu in September of 1944. This is a later production M4A2 with the simplified single piece glacis plate and larger hatches for the driver and co-driver. Unlike similar late M4A3 75mm tanks, these M4A2s retained the dry stowage ammunition racks in the sponsons and had appliqué plates added in these areas at the factory. (Naval History and Heritage Command)

Exhausted Marines rest in the shade of an M4A2 Sherman on Peleliu in September 1944. The Marines suffered numerous casualties from the tropical heat, which is reported to have reached 115° Fahrenheit in the daytime. This is a late-production tank with the large hull hatches, and high bustle turret with a loader's hatch. The hull rear has the folding storage rack but no spare track holders. (National Archives)

Large-Hatch M4A3 75mm

Numerous changes were incorporated into production of the M4 series, as has already been noted. Since it was of the utmost importance that no change in any way slow the flow of tanks rolling off assembly lines, most new components were designed as direct replacements for corresponding earlier parts.

Beginning production in February of 1944, the M4A3 underwent a major design change. The modifications were undertaken to address shortcomings with the original production types or to simplify and thus speed production. A new upper hull was constructed with a simple, single-piece front armor plate. This change saved numerous man-hours of welding. The thickness of the new frontal armor plate was also increased to 2½ inches and it was set at a steeper, 47° angle. The new hull had enlarged hatches for the driver and co-driver in a cast plate in the hull roof. The earlier smaller hatches had proven difficult to exit in an emergency.

The second major change that was incorporated at this time was wet ammunition stowage. Ammunition racks in the earlier tanks were fitted in the hull sponsons and were vulnerable to even minor penetrations. The new design moved the ammunition to the hull floor, where the shells were encased in racks filled with a water and anti-freeze solution. This arrangement was designed to reduce the likelihood of propellant fires being ignited by penetration. The use of under floor wet storage also eliminated the need for appliqué armor plates on the hull side.

This large-hatch M4A3 was photographed at the General Motors Proving Ground on 13 April 1944, only two months after the large-hatch variant was introduced. The tank is armed with a 75mm gun and features wet ammunition stowage, which lessened the risk of an ammo explosion. This particular tank was built by Fisher.

M4A3 Rear Deck Arrangement

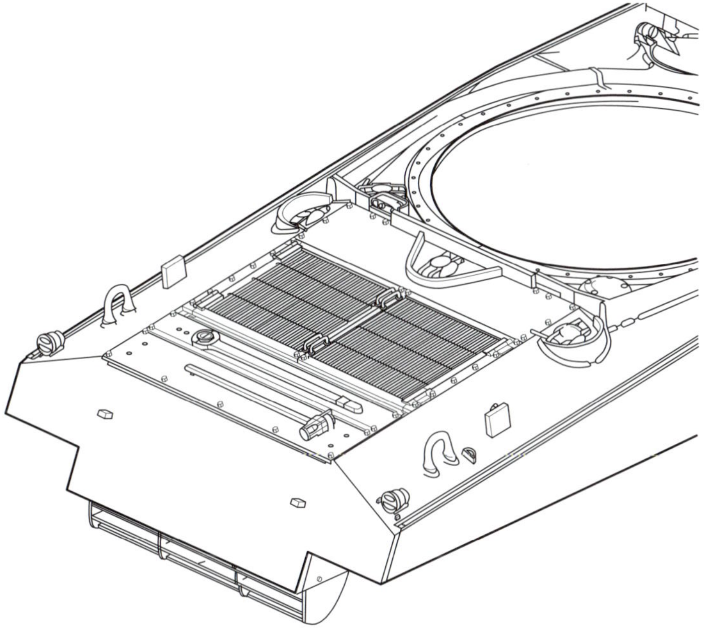

The Ford GAA V-8 engine powered the M4A3 Sherman. This water-cooled 60° gasoline engine displaced a massive 1,100 cubic inches (18.1 L) and produced 500 bhp. It was developed from an experimental Ford aircraft engine and became the U.S. Army's preferred tank power plant.

An M4A3 (75)W advances toward a burning vehicle across an open field near Lackenhauser, Germany, on 26 April 1945. The tank was part of the 8th Armored Division, which pushed to the Elbe as part of the XIX Corps. The tank's rear deck is piled with stowage, a common sight on U.S. vehicles in the last months of the war. (National Archives)

Even with the thicker armor of the new glacis and differential cover, the new M4A3s were still vulnerable to German anti-tank weapons. The addition of sandbags, often filled with concrete, was a common practice to gain additional protection. The tracks are fitted with the extended end connectors to improve flotation on soft ground. (Patton Museum)

A whitewashed M4A3 of 7th Armored Division supports troops from the 23rd Armored Infantry Battalion in the strategically important road and railway junction of St. Vith, in southeastern Belgium, in January of 1945. A brightly colored aerial recognition panel is displayed on the engine decks of the tank to identify it to friendly fighter-bombers. (Patton Museum)

This 7th Armored Division M4A3 displays all of the characteristics of a late-war Sherman. Sandbags have been added to the glacis and logs to the hull side for added standoff armor protection against the Panzerfaust. The tracks are fitted with extended end connectors. Sommerfield matting has been applied to the hull and turret sides for hanging foliage for camouflage. The vehicle is covered with the crew's personal gear. The turret is equipped with a vision cupola. (National Archives)

This M4A3 has struck a mine, resulting in damage to the right side track and drive sprocket. Tape has been laid in the area where mine clearing has begun. The vehicle awaits recovery and many external items are intact. The M2 HB .50-caliber machine gun remains on the turret rear under its canvas cover. No markings of any type are visible, not even the registration number on the rear hull side. (Patton Museum)

This M4A3 of the 6th Armored Division is undergoing repairs after being knocked out in the fighting to break out of the Bastogne area. A round has penetrated the upper edge of the hull side armor. The tank is equipped with a mix of road wheels including the late smooth-faced dished type, and there are extended end connectors on the tracks. Recovery and repair were essential for the success of U.S. armored operations in Europe. (Patton Museum)

M4A3s and a Jeep, assembled in a rear area, receive a fresh coat of whitewash. Various forms of winter camouflage were applied to vehicles in the winter of 1944-45. When sufficient white paint was unavailable, whitewash, a mixture of slacked lime and chalk, was used. At times chalk, snow, and bed linens were applied to make the dark Olive Drab vehicles less prominent in the snow. (Patton Museum)

A knocked-out M4A3 was an all too common sight in the drive into Germany. The small braces on the inside of the front fenders allowed crews to install a length of lumber, creating a shelf on the front glacis to carry various items. Here the crew has placed sandbags and spare road wheels for additional armor. (Patton Museum)

Late in the war U.S. forces faced very little armored opposition. The danger to tanks was in built-up areas from hand-held anti-tank weapons, the Panzerfaust, and Panzerschreck. Seldom are tanks seen at this time without infantry support. (Patton Museum)

3rd Armored Division tanks approach the spires of Cologne Cathedral on 6 March 1945. A pair of M4A3 (75)W tanks are accompanied by a combat veteran mid-production M4A1 in the center. All the tanks carry a large amount of stowage on their engine decks. (National Archives)

According to the U.S. Army Signal Corps, these soldiers of the 16th Infantry Regiment, 1st Infantry Division are riding their M4A3 Sherman into Schopen, Belgium, on 21 January 1945. The tank is whitewashed and carries a fresh covering of snow. Most of the tank riders have winter camouflage clothing. (National Archives)

Finally the 75mm turret also was transformed. The overhang, which housed the radio at the rear of the turret, was raised to allow the new, larger hatches to open even when the turret was traversed to the rear. These new turrets are often referred to as "high-bustle" models. At the same time that the design was changed, the pistol port was re-introduced to the left side wall. The loader also was furnished with a hatch in the turret roof above his position. No longer would this crewman have to go under the gun to exit out of the commander's hatch in an emergency. Stowage brackets for the .50-caliber machine gun were added to the turret rear.

Many smaller features also began to appear at this time of transition. Gun travel locks had become standard. The rear plate of the upper hull received a stowage rack and brackets for spare track. All periscopes had a protective guard, the gunner's periscope received an armored cover. The siren was relocated to the glacis armor behind a brush guard.

Total production of the M4A3 (75)W reached 3,071 by March of 1945. These tanks served alongside 76mm tanks until the end of the fighting in Europe. Many of the final tanks to reach Marine Corps units for the final battles on Iwo Jima and Okinawa were of this type.

D47366 early flame cut sprocket

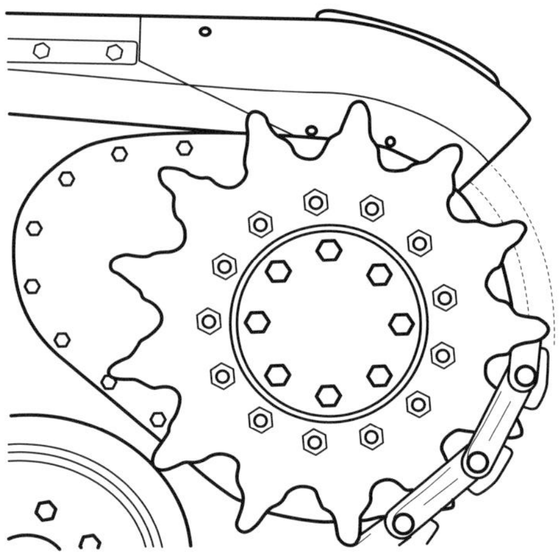

Flame-cut drive sprocket

The most common of the M4A3 series was the large-hatch hull, wet-stowage 75mm tank or M4A3 (75)W. This knocked-out example displays the large oval hatches for the driver and co-driver. The ammunition storage was moved to the floor of the lower hull in lined, water-filled racks that reduced the likelihood of fire from a penetration. The appliqué plates on the hull sides to protect the ammunition racks were no longer necessary.

The M4A3 (75)W began to appear on the European battlefield in the fall of 1944. It differed from the M4 by having its simplified front glacis plate at a steeper 47° angle, larger hull hatches in the roof armor, and later-production high bustle turrets with an oval hatch for the loader. The rear hull armor plate extended down to protect the radiators in the rear of the engine compartment. (National Archives)

Coed, a Marine 4th Tank Battalion flame tank, fires its weapon on Iwo Jima during February 1945. The tanks on Iwo Jima have many details specific to the battle. Large pieces of lumber cover the upper hull sides, spaced out from the hull to allow a layer of concrete to be poured between. A telephone box is installed on the left side spare track brackets. The cylinder on the engine deck is a fuel tank from an M3 Stuart light tank used for fuel for the flamethrower. (National Archives)

A Marine Corps M4A3 on Okinawa in June 1945 has been covered with track links for additional armor protection. The track links leave very little room for visible markings other than a tactical number at the upper edge of the turret. The tank has an all-round vision cupola that has been installed in a position which makes the hatch lid open forward. (National Archives)

A roadway bridge has collapsed under the terrific weight of this 6th Tank Battalion M4A3 on Okinawa. The tank has track links welded to the turret for extra protection, and extended end connectors on its tracks. The vision cupola here is turned to allow the hatch to open forward, perhaps to give the commander some measure of protection while operating with his head out of the hatch. There is an air-recognition star with bars on the turret roof. (National Archives)

The crew of this Marine Corps M4A3 set up residence on Okinawa and harvest local vegetables. This rare color shot provides great insight into the details of tanks of the period. Track links have been used to add additional protection. Small patches of armor plate have been welded to the hull side. Large amounts of personal gear are covered by camouflage tarps on the rear engine deck. Light-colored, dried mud is caked into the suspension. (National Archives)

Large-Hatch M4A3 (76)W

Chrysler's Detroit Tank Arsenal began production of the 76mm-armed M4A3 in March of 1944. The new vehicle incorporated an updated hull with larger hatches for the driver and co-driver, and a simplified one-piece glacis armor plate. The hull and drive train were identical to the 75mm-armed version with the exception of the ammunition racks. The ammunition stowage was enclosed in water tanks, and moved from the hull sponsons to reduce the risk of fire if the hull was penetrated.

Initially the tanks were produced with the smooth-barrel M1A1 76mm gun. Later variants received the M1A1C or M1A2 gun with provisions for adding a muzzle brake. Another feature that began to appear on later-production turrets was provision for mounting a canvas dust cover to seal the openings around the outer gun rotor shield.

The 76mm gun provided approximately one more inch of armor penetration than the 75mm gun offered at comparable range. The high-explosive round was inferior to that of the 75, so production of the 75mm M4A3 was continued. Not until sufficient supplies of tungsten-core HVAP ammunition became available was the 76mm gun capable of dealing with the frontal armor of the Panther at battlefield ranges.

The M4A3 (76)W tanks began to arrive with the armored units in the fall of 1944. They followed the M4A1 (76) into service as there was need to establish the logistical supply of spare parts for the new Ford engine that had not yet been used in Europe. But by the later stages of the war in Europe, the M4A3s were the most common tanks in the American Armored Divisions and Independent Tank Battalions.

Both the Chrysler Tank Arsenal and General Motors' Fisher Body assembly plant made the 76mm M4A3, turning out a total of 4,542 units by the end of production in April 1945. The first 1,400 tanks built by Chrysler, and all 525 of the M4A3 (76)s made by Fisher used the narrow track and original style of vertical volute spring suspension.

Like the 76mm-armed versions of the M4A1, the M4A3 (76)W featured a turret derived from that developed for the ill-fated T23 medium tank program. (Patton Museum)

The M4A3 (76) first appeared in combat with the 1st Armored Division in Italy. This 1st Tank Battalion tank was photographed near Pauglia on 1 September 1944. The division used colored stripes on the barrels of its tanks to denote the platoon and company by number and color of the stripes. The hull side star has been over painted in green, and there is a pin-up girl taped to the side of the hull just forward of the star. The instrument beside the commander is a vane sight added for use in indirect fire missions, a common use for tanks in the Italian campaign. (National Archives)

The crew of this new M4A3 (76) are loading ammunition near Pisa, Italy, on 19 August 1944. The 1st Armored Division was the first to receive the new tanks for combat. An indirect fire sight is mounted near the gunner's periscope. Tanks were often used as artillery in the hilly terrain of Italy. (National Archives)

Crews of 1st Armored Division tanks familiarize themselves with their new mounts. The M4A3 hull is identified from the rear by the hull armor extending down to protect the radiators above the exhaust, and the sheet metal exhaust deflector below. The new tanks have spare track racks on the corners of the rear hull, and a folding stowage shelf between them. (National Archives)

The crews of the new 1st Armored Division tanks practice gunnery, adjusting to the new 76mm gun. The tanks display areas of fresh paint covering the stars on the hull sides and platoon number on the turret rear. The vehicle in the foreground carries a fuel can in the left side spare track bracket. The tracks are all the T48 rubber-chevron type. (National Archives)

The crew of an M31 ARV is preparing this M4A3 (76) for recovery in Italy on 19 September 1944. The rubber portions of the track and road wheels have burned, scorching the paint on the hull sides and sand shields. The turret is the early type with the split-hatch cupola for the loader. (National Archives) (National Archives)

Blondie, an M4A3 (76) of the 751st Tank Battalion, passes a flock of sheep near Porretta, Italy on 31 December 1944. The tank is equipped with T48 rubber-chevron tracks with extended end connectors. The star on the hull side has been painted out, while the turret star has been obscured with a coating of mud. The two hoops ahead of the commander's cupola protect an added sight for indirect fire. (National Archives)

A 14th Armored Division M4A3 (76) practices artillery type indirect fire in Huttendorf, France, on 11 February 1945. The 14th constructed elaborate racks for sandbags on its tanks' hulls and turrets to serve as supplemental armor. Here the hull racks have not yet been filled. Bumper code markings which were covered by the sandbags were repainted onto the gun barrel. The turret and barrel have been camouflaged with a darker color applied with a brush. (National Archives)

From the Battle of the Bulge through the end of the war in Europe 76mm M4A3 tanks were a common sight. Here a 3rd Armored Division tank from "C" Company of the 774th Tank Battalion moves past a knocked out Panther Ausf. G on 17 January 1945. A star with surround is visible below the layer of sandbags and snow on the frontal armor. (National Archives)

An M4A3 (76) of the 761st Tank Battalion "Black Panthers" appears during the late winter of 1945. The troops of the 761st served gallantly throughout the drive into Germany. The tank has a small number of sandbags on the hull front, and extended end connectors on steel chevron tracks. A canvas cover protects the front machine gun. (Patton Museum)

A 5th Army M4A3 (76) negotiates the soft soil of a hillside near Monterumici, Italy, on 17 April 1945. The extended end connectors appear to be of little help on this surface. The use of supplemental armor was far less common in the Italian fighting, and in the later stages of the war, tank-to-tank fighting became less frequent. (National Archives)

M4A3s with African-American crewmen pass through the ruins of a German town. Typical of U.S. tanks of this period, the front armor plates are covered with supplemental sandbag protection. All vehicles retain the national star marking on the turret yet only some shipping stencils remain on the hull of the forward tank. (Patton Museum)

An M4A3 of an unknown unit is recovered in the snow during the fighting following the Battle of the Bulge. The left side track is missing and the tank has left the road, suggesting damage from an anti-tank mine. The winter camouflage was either hastily applied or is well worn. (Patton Museum)

Early 76mm-armed Turret

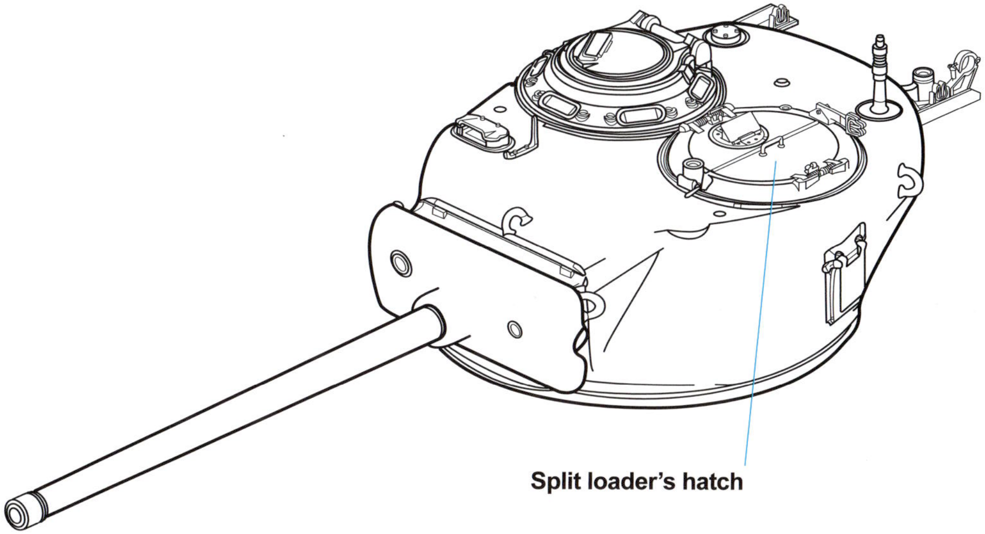

This M4A3(76)W was attached to the 19th Tank Battalion, 9th Armored Division, while the unit operated in Belgium in November 1944. It features two-tone green camouflage with black splotches and is devoid of the large white star typically found on allied vehicles.

A pair of M4A3 Shermans halt in the ruined streets of the Surbourg area of Alsace on 19 January 1945. The front tank is a 76mm-armed tank, the rear a 75mm version. Both tanks have nets and Hessian tape strung over the turrets and hull to break up their silhouette. The forward tank has sandbag armor on the front hull beneath the nets. The .50-caliber machine gun is mounted on the loader's split-hatch cupola. The tracks are T54E1 steel-chevron type with extended end connectors on the forward tank. (Patton Museum)

M4A3 tanks of an unknown unit advance along a rain-slicked road in Germany. The tanks appear to be crewed by African Americans in the original print. The forward tank displays a disruptive camouflage pattern, possibly in a lighter green than the Olive Drab base paint. The gun is the M1A2 76mm with a collar to protect threads for a muzzle brake. (Department of Defense)

The crew of this M4A3 (76) have time to relax at war's end. The tank has an extensive set of brackets attached to the hull side and turret for carrying supplemental sandbag armor. The sandbags added greatly to the weight of the tank and there were concerns about the effect on automotive reliability. The tank is parked in the midst of a marked cycling path. (Patton Museum)

A line of 8th Armored Division tanks take a break beside houses in the village of Bergenhausen in March of 1945. The crew of the forward tank, an M4A3 (76), has used anything at their disposal to add some bit of stand off protection from shaped-charge weapons. The hull front is covered with sandbags and track grousers, the hull side has a large rectangle of either lumber or armor plate, and the turret is lined with narrow planking, giving the vehicle a distinctive vagabond-like appearance. (Patton Museum)

A 7th Armored Division M4A3 (76) passes a burned-out civilian automobile on the outskirts of Haina, Germany, on 30 March 1945. Sandbags cover the tank's front plate and evergreen boughs camouflage the hull, turret, and gun barrel. Ahead of the driver's position on the top edge of the front armor plate is painted *Minnesota Kid.* This is probably a nickname of the driver rather than the tank, as there is another illegible name on the turret armor hidden by the gun muzzle. The background vehicle is an M4A3E8 (76) with the 23-inch wide T66 track. (National Archives)

M4A3 (105)

The newly designed 105mm howitzer in the M52 combination mount was also installed on the M4A3 tank chassis. The M4A3 with the Ford GAA gasoline V-8 engine was the tank preferred by U.S. units. The tanks were built utilizing the new, lager hatch and simplified glacis armor with steeper 47° inclination. The 105 variant retained dry ammunition stowage for 66 rounds. No appliqué plates were installed on the hull sides.

Production began at the Chrysler Tank Arsenal in May of 1944. By September the initial 500 tanks had been produced with the VVSS and narrow track. Production transitioned to the new horizontal volute spring suspension. Output of the M4A3 (105) totaled 3,039 tanks by the time production ceased in June of 1945. All 105mm-howitzer tanks were built at the Chrysler facility.

M4A3 (105) tanks began to reach the European battlefield in the fall of 1944. Initial tanks were equipped with the split-hatch cupola for the commander. This cupola was later replaced with the vision cupola as more of these became available. The tanks were attached to headquarters companies of the U.S. tank units and served as the artillery element. Later, toward the end of the campaign, the M4A3E8 (105)s with the new wide track and HVSS began to appear in Europe, but photographs of these vehicles are rare.

The tanks are frequently seen towing ammunition trailers because they could use up tremendous stocks of ordnance when functioning as artillery pieces.

A new M4A3 (105) tank displays all the factory fittings such as sand shields and the dust cover around the gun mount. All of the M4A3s with the vertical volute spring suspension and 105mm howitzer were produced by the Chrysler Tank Arsenal between May and September of 1944. (Patton Museum)

New tanks arrive at the port city of Cherbourg on 3 November 1944. The howitzer tank has been protected for the sea voyage by sealing all hatches and joints. The dark compound on the suspension was a sealant to keep the lubricated parts from being damaged by salt water. The box on the engine deck carries all the tank's factory-supplied accessory parts. (National Archives)

105mm Howitzer and Rotor Shield

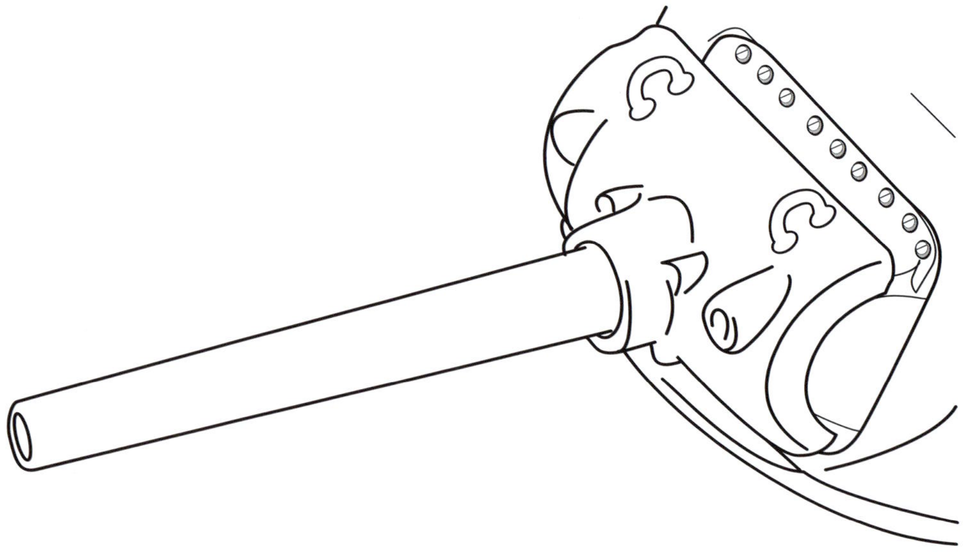

A pair of white camouflaged 14th Armored Divison M4A3 (105) tanks advance in the area of Hochdelden, France during late January 1945. The tanks carry sandbags on their front plates covered with nets and snow. The stowage on the rear deck has been covered with white cloth. Too large for local bed linens, this cover is possibly a tarpaulin painted white. (National Archives)

Crews load of a 730th Tank Battalion M4A3 (105) with ammo in a street near Düsseldorf, Germany, on 3 March 1945. The tank is equipped with the all-round vision cupola for the commander, and T49 three-bar steel cleat tracks. The hull has a thorough covering of sandbags and logs for added armor protection. (National Archives)

Tanks used as artillery generally fired considerably more ammunition than when used in other roles. Typically Sherman 105 howitzer positions were surrounded by piles of ammunition, spent casings, ammunition boxes, and packing tubes. The howitzer tanks were attached to headquarters companies and provided the artillery element for the armored units. (Patton Museum)

M4A3E2

Anticipating the need for an assault tank to support infantry during the coming invasion of France, Ordnance developed the M4A3E2. It was found that with minor modifications, the new 47°-hull M4A3 could carry significantly greater armor protection.

First, 1½-inch armor plates were added to the hull glacis and side armor, bringing the thickness to 4 and 3 inches, respectively. The differential/final drive cover was replaced with a new casting with a thickness of up to 5½ inches. The new turret was based on the shape of the new 76mm turret that had been developed in the T23 program, but the pistol port was eliminated from the left side wall. The massive casting had armor 6 inches thick on the sides and 2½ inches in the rear, with the roof plate welded in place.

The original concept of an infantry support tank determined that the M3 75mm gun would be used in a mount derived from the M62 mount for the 76mm turrets. The outer gun shield was produced by adding additional plate to the outside of the standard 76mm part, bringing the thickness up to a total of 7 inches.

The only automotive change was an increase of the final drive ratio in an attempt to extend power-train longevity. This change cut the top speed to around 22 m.p.h. All the tanks were built with the extended end connectors on the tracks to reduce ground pressure from the massive 84,000-pound weight. The increased weight overburdened the tanks' standard VVSS suspension. Period photos often show a mix of road wheels, the result of replacement following bearing failures, and compressed suspensions.

The Fisher Tank Arsenal produced a total of 254 of the M4A3E2s, which began to arrive with combat units in the fall of 1944. Dubbed the "Jumbo" because of its massive size, the tank could absorb punishment from German guns, and was a very popular point tank with U.S. armored units until the end of the war in Europe.

M4A3E2 Up-armored Turret

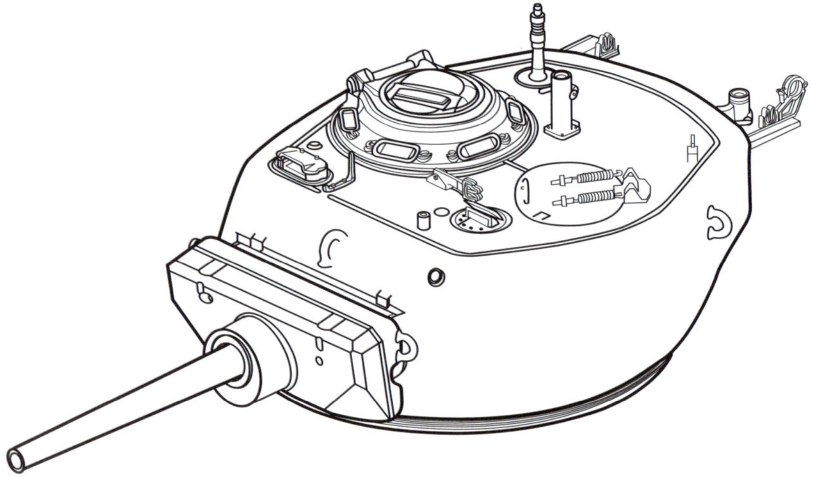

The M4A3E2 "Jumbo" carries massive armored protection when compared to the standard M4 series. This factory photo from June 1944 provides details of the huge turret and differential cover castings, and the welds at the hull corners attaching the additional armor plate. All 254 tanks were built at General Motors' Fisher Tank Arsenal. (Patton Museum)

A thick-skinned M4A3E2 Assault Tank moves through the streets of Köln ("Cologne"), Germany. Although not needed on the hard-surfaced city streets, all M4A3E2 tanks were equipped at the factory with the extended end track connectors, popularly known as "duck bills," which kept ground pressure at approximately 14 psi. (Patton Museum)

This M4A3E2 of the 4th Armored Division near the city of Koblenz (“Coblenz”), Germany, on 9 March 1945 has been rearmed with the 76mm gun. Though originally an assault tank fitted with the 75mm gun, many of the vehicles were up-gunned to increase their armor-piercing performance for tank-to-tank fighting. The .50-caliber Browning machine gun has been relocated ahead of the loader’s hatch and a .30-caliber added for the commander. (National Archives)

The “Jumbo” was heavily armored but not impervious to all weapons. This 6th Armored Division tank was knocked out by a Panzerschreck in on 27 March 1945. It has also been rearmed with the more powerful 76mm gun. Note that the rubber portion of the right side track has burned. (Patton Museum)

Infantrymen ride aboard this 2nd Battalion, 3rd Armored Division “Jumbo” on 3 January 1945. The tank retains its 75mm gun and is equipped with the standard welded road wheels. The bearings of these wheels were prone to failure due to the weight, and are often seen replaced by later welded wheels and dished wheels both with the larger wheel bearings. (Patton Museum)

A 9th Army “Jumbo” passes a group of G.I.s and a nun in the Gustorf area, near Düsseldorf, in the spring of 1945. The unit number has been removed by the censor. This tank also retains the 75mm armament. All M4A3E2s were equipped with the extended end connectors on the tracks. The canvas cover is installed on the front machine gun. To the left of the road are a U.S. M3 half track and Jeep. (Patton Museum)

This "Jumbo" of the 4th Armored Division seen in Alzey, Germany, on 20 March 1945, has been rearmed with a 76mm M1A1 gun. The sides of the tank are covered by Sommerfield matting for attaching foliage as camouflage. The relocation of the .50-caliber machine gun to the loader's position made the weapon more accessible for use against German troops wielding hand-held anti-tank weapons. A .30-caliber Browning has also been added to the turret roof ahead of the commander. (National Archives)

This factory overhead shot shows to good advantage the thickness of the massive turret casting and gun shield, and the welds attaching the additional hull armor. The additional armor of the M4A3E2 increased its weight by over 11,000 pounds in comparison to a standard M4A3. The tank is still fitted with the factory-installed sand shields that were often removed in combat. (Patton Museum)

This "Jumbo," seen on 3 January 1945, has received a fresh coat of winter white camouflage paint. The M4A3E2's short fender on the lower hull side allowed the sand shields to be installed over the additional width of the tracks with extended end connectors. The sand shields interfered with maintenance and were often removed by units after shipping. (Patton Museum)

An M4A3E2 of the 737th Tank Battalion drives at high speed through Luxembourg on 22 January 1945. The 737th supported the 5th Infantry Division on the southern flank of the Ardennes. To extend the life of the automotive components of the much heavier Jumbo, the gear ratio of the final drive was increased. This reduced the maximum speed of the tank from 26 m.p.h. for a standard M4A3 to about 22 m.p.h. on the Jumbo. (Patton Museum)

A 4th Armored Division M4A3E2 passes a column of burning German vehicles on 20 March 1945. The additional armor of the “Jumbo” made it more capable than any U.S. tank of absorbing punishment from German tank guns. The second Sherman in the column has had additional armor added to the glacis, differential cover, and turret sides to augment the protection similar to the “Jumbo's.” (National Archives)

A re-manufactured M4A3 tank fires its front-mounted flame gun in training at Ft. Benning, Georgia, in May of 1945. The E4-5 flame gun replaced the .30-caliber front machine gun and carried a fuel tank in the right sponson. The guns' limited range and fuel supply made them less than satisfactory. (National Archives)

The Sherman Duplex Drive tank was the solution for making the M4 series amphibious. A limited number were built from M4A1, M4A2, and M4A4 tanks with the addition of a collapsible canvas screen supported by pneumatic bags. The tank was propelled in the water by two propellers powered by linkages to the drive train. (Patton Museum)

M4 Specialized Use

As the war progressed and the M4 series reached full production, a myriad of special-purpose vehicles and attachments were developed for specific roles to support the armored units. It is only logical that many of these vehicles would be based on the Sherman to simplify logistical supply for systems common with the tanks already in service. In keeping with the scope of this volume we will take a brief look at types that allowed the vehicles to continue to be used as gun tanks in U.S. service.

Amphibious

The desire to be able to land tanks along with the first wave of infantry during amphibious assaults led to the development of the Duplex Drive Sherman. The tanks were fitted with a waterproof canvas screen that was held erect by pneumatic bladders. The tank was propelled in the water by two 26-inch propellers. Driven by the tracks, the propellers pivoted up when not in use. The U.S. converted 350 M4A1 (75) tanks before the D-Day landings, while the British conversions were based on M4A2s, and M4A4s. A majority of the U.S. tanks launched on June 6th at some distance from the shore were lost in the rough seas. DD tanks were later employed in river crossings in the European campaign, including the crossing of the Rhine River in March of 1945.

Bulldozer Tanks

Work began in 1942 to develop an earth-moving capability for the M4 series. Initial V-shaped blades were relatively ineffective. The standardized M1 tank dozer had comparable capabilities to a Caterpillar D-8. The units mounted to the second bogie station along the side of the tank. A housing carrying the hydraulic cylinder was mounted to the final drive cover. The unit converted any standard M4 series tank, and nearly 2,000 of the various models were manufactured during the war.

Flamethrowers

Several attempts were made to integrate flame-throwing tanks into U.S. armored units. These included front-mounted models with limited range and fuel supply, as well as weapons that replaced the tank's main gun in the turret. Flamethrowers were unpopular with the tank crews, who feared for their safety while carrying the flammable fuel supply. A few of the small front-mounted units saw service in Europe, but flame tanks were used extensively in the Pacific campaign until the end of the conflict.

Mine-Exploding Vehicles

Various attempts were made to attach mine-clearing devices to the Sherman. The U.S. tried mine plows and copies of the British flail system with a motor-drive drum ahead of the vehicle that turned with lengths of heavy chain exploding the mines in the tank's path. The solution was found in several designs that mounted huge indestructible rollers ahead of the vehicle. The type used on gun tanks was the T1E3 pictured on page 73.

Rocket launchers

Ordnance developed a series of rocket-launching mounts that attached to the turret of the Sherman. The most common of these was the T34 and T34E1 "Calliope" which carried 60 rockets in a jettisonable arrangement of tubes above the turret. These weapons provided a concentrated barrage of fire a short distance ahead of an armored assault.

Duplex Drive Shermans come ashore from an LST and collapse their flotation screens during the invasion of Southern France on 15 August 1944. Many crews removed the canvas screens once the tanks were inland and were called upon to be used as normal gun tanks. Only the tanks based on the M4A1 were used in combat by U.S. troops. (Naval History and Heritage Command)

A T1 Mine Exploder mounted to the front of an M4A1 Sherman. The nickname *Aunt Jemima* comes from the resemblance of the huge rollers to pancakes. The rollers were powered by a chain driven by the track drive sprockets. The large plate at the rear of the vehicle allows another tank to push the vehicle if stuck in soft ground. A number of types of mine exploders were manufactured for use with the M4 series. (Patton Museum)

An M4 Sherman with the M1 Tank Dozer attachment in France during the summer of 1944. Although the dozer blades were constructed of mild steel rather than armor, they did provide a bit of additional frontal protection. They are often seen used as point tanks at the front of armored columns. The dozer kit attached to the center bogie on each side, and the hydraulic cylinder was mounted to the tow lugs of the differential cover. Nearly 2,000 tank dozer conversion kits were manufactured during WWII. (Patton Museum)

The T40 rocket launcher, which housed and fired 20 7.2-inch rockets, first saw combat in the Mediterranean Theater. The launchers, later standardized as the M17, were fitted on four tanks of each of the 752nd Tank Battalion and the 760th Tank Battalion and were test fired near Scansano in August 1944. The T40's performance was evidently less than stellar, as by October the 752nd began to re-equip with the 60-tube, 4.5-inch T34 rocket launcher. All eight T40-armed M4A1s were rearmed with the T34 on 26 December 1944.

The most common type of rocket launching system mounted on the Sherman was the T34. It was named “Calliope” for its resemblance to organ pipes, and the sound it made when fired. The launcher carried 60 M8 4.5-inch rockets and could be jettisoned once fired. The elevation of initial models was controlled by a rod clamped to the main gun, which could not be fired until the clamp was removed. This made the weapon very unpopular with crews until modifications were made. (National Archives)

A U.S. unit trains for amphibious landing operations in 1944. The tank is an M4A4 with an M1 Tank Dozer attachment. The forward hydraulic cylinder was protected by a large assembly of rectangular armored plates. The M4A4 was not used by U.S. troops in the ETO, and this tank appears to have a British registration number painted out on the hull side. The earth yellow camouflage paint indicates that the tank may be a veteran of the Italian campaign. (Naval History and Heritage Command)

M4A6

In the quest for effective alternative tank power plants Caterpillar Tractor Company modified a Wright radial aircraft engine to operate on diesel fuel. The engine would also run on a number of other fuels, including 100-octane gasoline. The engine was tested in the M4E1 and production of the M4A6 was authorized with the production Caterpillar RD1820 engine. To accommodate the large engine, the tank's hull was lengthened to the same dimensions as those of the M4A4, and large cast bulges were installed in the engine deck and hull floor plates for clearance. The tanks were all built with a cast section for the glacis and hull hatches like the M4 composite. The new larger hatches were used for the driver and co-driver.

Chrysler began production of the M4A6 in October of 1943. The tanks were equipped with the late, sharp-nosed differential/final drive unit. The T48 rubber chevron tracks are seen in all production photographs. Even in such a small production run there is variation in the turrets used. Some tanks were built with the late, low bustle turret that lacked the pistol port, while others received the new high-bustle turret with the hatch for the loader in the left roof. The hull was equipped with appliqué plates on the side armor, and dry ammunition stowage was retained.

Production was halted in February of 1944 after only 75 tanks were completed. The tanks served in training units and never saw combat, but the engine represents an early step toward diesel and multi-fuel power plants for U.S. tanks.

The first Chrysler-produced M4A6 is loaded and prepared for rail transport. The M4A6 can be identified by its longer hull and by its cast forward section, which was identical to the M4 composite-hull tanks. This tank's turret is the late low-bustle style that had no loader's hatch or pistol port in the left turret wall. Some of the M4A6s produced received the later high-bustle design. (Patton Museum)

An M4A6 parks at the Chrysler plant. Chrysler produced 75 of these M4A6 tanks with the Caterpillar RD1820 Radial Diesel engine between October 1943 and February 1944. The engine's large size required the use of the longer hull length of the M4A4 series. While the tanks did not see combat, they were used in training. (Patton Museum)

In the engine deck of the M4A6 is another of the tank's distinguishing details: a large domed plate that was necessary to cover the large engine. The tank had a similar bulge in the floor plate. The M4A6 did not prove a successful design but provided an early step in the development of purpose-built diesel power plants for U.S. tanks. There is no pistol port in the turret. (Patton Museum)

The M4A3E8 represents the ultimate development of the M4 Sherman series. The tank incorporated all the improvements developed throughout the entire production. This is an early vehicle photographed on 20 September 1944. It is from the first batch and retains the split hatch cupola for the loader. The tracks are the single-pin T66 type. (TACOM LCMC History Office)

At the Chrysler Tank Arsenal in Detroit, 76mm and 105 Howitzer tanks undergo final assembly. This massive building stands as testimony to the manufacturing might of the United States economy during the war years. All of the 76mm tanks have the late turret with the oval loader's hatch and muzzle brakes. (Patton Museum)

M4A3E8

Throughout 1943 specialists strove to improve the automotive performance of the M4 series. Earlier efforts had yielded a system that used the standard narrow track with a new suspension that had horizontal volute springs and shock absorbers on the first and last bogie station. These results spurred work toward a new HVSS system with dual wheels and a wider track to reduce ground pressure and improve cross-country performance.

In March of 1944 production began of the initial 500 M4A3s with 76mm gun equipped with the new suspension system. The new tanks were designated M4A3E8, 2,617 being produced by Chrysler by the end of WWII. When they reached the troops, the ride was so greatly improved over the older models that the nickname "Easy Eight" soon stuck. The tanks were initially outfitted with the 23-inch wide single pin all steel track T66, and later with the T80, a dual pin steel design with a steel chevron shaped cuff. The wider track increased the contact area and reduced ground pressure from over 14 psi to about 11.

The new tanks were well received by ETO armored units in early 1945. Besides having improved flotation, the new tanks incorporated all of the updates found on the 76mm-gunned tanks, most featuring the oval loader's hatch and muzzle brakes on the gun.

The suspension was so successful that all of the M4 series production began to receive it in late 1944. The M4 (105), M4A1 (76)W, M4A2 (76)W, M4A3 (75)W, M4A3 (76) W, and M4A3 (105) all received the new tracks and suspension, and many other vehicles were converted after the end of the war. The M4A3E8 (76) and (105) tanks were the preferred types used for service with the U.S. Army in the immediate post-war period and into the Korean War.

Late 76mm-armed Turret

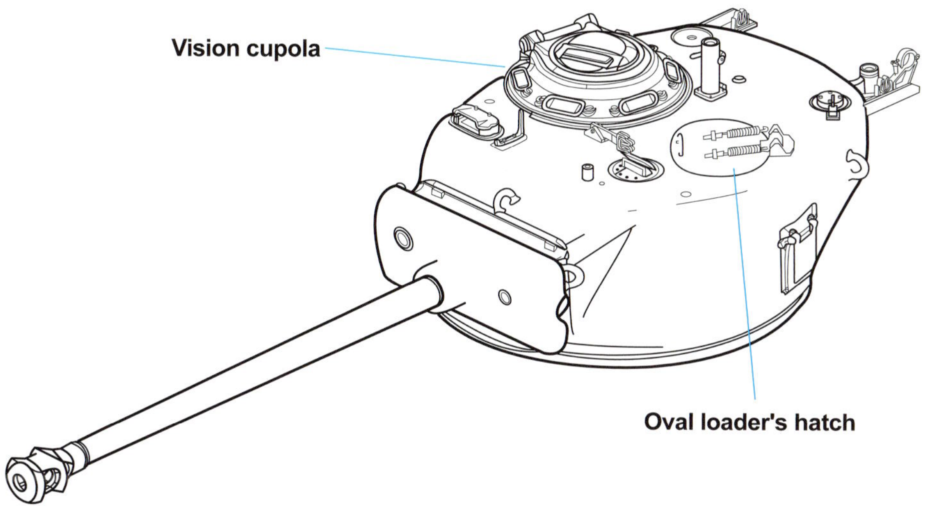

M4A3E8 tanks of the 745th Tank Battalion pause in a battle-scarred village in Germany on 1 April1945. The stars have been overpainted on the forward two vehicles. All three tanks are early-production Easy Eights retaining the split-hatch rotating cupola for the loader, and single-pin T66 tracks. The vehicles appear to be recently issued with very little stowage and no visible damage to the fenders. (National Archives)

An M4A3E8 with pine boughs as camouflage advances past a burning structure. The tank is equipped with a muzzle brake, which was developed for the U.S. 76mm gun by copying examples of German brakes used on Panzer IVs. The brake reduced the recoil of the gun, saving wear and tear. It also directed the muzzle blast to the side, reducing the amount of debris disturbed on the ground. This made it easier for a gunner to lay the gun on a new target after a round had been fired. (Patton Museum)

Once the 14th Armored Division received their new M4A3E8 tanks, they continued their practice of adding elaborate racks for adding sandbag protection to the tank. The addition of HVSS and the T66 track added an additional 2,950 pounds over a standard M4A3; with the weight of the sandbags the performance of these tanks must have been greatly reduced. Divisional markings have been repainted on the gun barrel. (National Archives)

An M4A3E8 (76)W of an unknown unit passes a burning German dwelling. The tank has a muzzle brake and T66 tracks. The turret is the later type with oval loader's hatch. The .50-cal. machine-gun mount has been relocated forward for handy use against positions that might hide an infantryman with a Panzerfaust. Hand-held weapons were the greatest threat to U.S. armor at this late stage of the war. (National Archives)

A column of the 781st Tank Battalion enters Bitche, France, in support of the 100th Infantry Division on 16 March 1945. The M4A3 has all the later-production details: an oval hatch for the loader, a muzzle brake, and T80 dual-pin live tracks. The tank is thoroughly covered with racks of sandbags on both the hull and turret and wide bands of a disruptive camouflage, probably black, have been sprayed over the Olive Drab paint. (National Archives)

A column of the 23rd Tank Battalion, 12th Armored Division stops along a German road on 20 March 1945 awaiting repairs to an M4A3E8 Sherman. The tank has a muzzle brake and T66 tracks. The cause of the breakdown is unknown but is not from enemy fire, or the column would not have stopped in such an exposed position. (National Archives)

Occupying armies meet on a road on the outskirts of Berlin, July 1945. This M4A3E8 of the 2nd Armored Division is likely a vehicle delivered after the war's end. It displays a complete set of markings and the name *Daytona Beach* on the hull side. The tank has a non-standard rack for personal gear on the turret rear and an antenna mounted in the bracket beside the co-driver's hatch in the roof of the hull. (National Archives)

Scores of M4A3E8 (75)W tanks are stockpiled in an Ordnance Depot in Manila, awaiting the invasion of Japan. These tanks never saw combat and many were converted to 76mm tanks post war. The new HVSS suspension systems also went into 75mm- and 105mm-M4A3s. The Sherman faced very little armored opposition in the Pacific, where the H.E. round of the 75mm tanks was preferred to the 76. (National Archives)

The Autobahn was one of the greatest achievements of the German Reich, yet here it serves the invading armies as a tool of conquest. The M4 Sherman is often regarded as having been inferior to its German adversaries, but it was part of the vast array of robust, reliable vehicles that the economic and manufacturing might of the United States developed and produced in huge numbers. (National Archives)